Springer Topics in Signal Processing

Volume 2

Series Editors

J. Benesty, Montreal, QC, Canada
W. Kellermann, Erlangen, Germany

Springer Topics in Signal Processing

Edited by J. Benesty and W. Kellermann

Vol. 1: Benesty, J.; Chen, J.; Huang, Y.
Microphone Array Signal Processing
250 p. 2008 [978-3-540-78611-5]

Vol. 2: Benesty, J.; Chen, J.; Huang, Y.; Cohen, I.
Noise Reduction in Speech Processing
240 p. 2009 [978-3-642-00295-3]

Jacob Benesty · Jingdong Chen · Yiteng Huang · Israel Cohen

Noise Reduction in Speech Processing

Prof. Dr. Jacob Benesty
Université de Quebec
Inst. National de la Recherche
  Scientifique (INRS)
800 de la Gauchetiere Ouest
Montreal QC H5A 1K6
Canada
benesty@emt.inrs.ca

Yiteng Huang
9 Sylvan Dr.
Bridgewater NY 08807
USA
ardenhuang@gmail.com

Jingdong Chen
Lucent Technologies
Bell Laboratories
600-700 Mountain Avenue
Murray Hill NJ 07974-0636
USA
jingdong@research.bell-labs.com

Prof. Israel Cohen
Technion - Israel Institute of
  Technology
Dept. Electrical Engineering
32000 Haifa
Technion City
Israel
icohen@ee.technion.ac.il

ISBN 978-3-642-00295-3   e-ISBN 978-3-642-00296-0
DOI 10.1007/978-3-642-00296-0
Springer Dordrecht Heidelberg London New York

Springer Topics in Signal Processing ISSN 1866-2609

Library of Congress Control Number: 2009921819

© Springer-Verlag Berlin Heidelberg 2009
This work is subject to copyright. All rights are reserved, whether the whole or part of the material is concerned, specifically the rights of translation, reprinting, reuse of illustrations, recitation, broadcasting, reproduction on microfilm or in any other way, and storage in data banks. Duplication of this publication or parts thereof is permitted only under the provisions of the German Copyright Law of September 9, 1965, in its current version, and permission for use must always be obtained from Springer. Violations are liable to prosecution under the German Copyright Law.
The use of general descriptive names, registered names, trademarks, etc. in this publication does not imply, even in the absence of a specific statement, that such names are exempt from the relevant protective laws and regulations and therefore free for general use.

Cover design: WMXDesign GmbH, Heidelberg

Printed on acid-free paper

Springer is part of Springer Science+Business Media (springer.com)

Preface

Noise is everywhere and in most applications that are related to audio and speech, such as human-machine interfaces, hands-free communications, voice over IP (VoIP), hearing aids, teleconferencing/telepresence/telecollaboration systems, and so many others, the signal of interest (usually speech) that is picked up by a microphone is generally contaminated by noise. As a result, the microphone signal has to be cleaned up with digital signal processing tools before it is stored, analyzed, transmitted, or played out. This cleaning process is often called noise reduction and this topic has attracted a considerable amount of research and engineering attention for several decades. One of the objectives of this book is to present in a common framework an overview of the state of the art of noise reduction algorithms in the single-channel (one microphone) case. The focus is on the most useful approaches, i.e., filtering techniques (in different domains) and spectral enhancement methods. Readers who are interested in algorithms based on multiple microphones (or microphone arrays) are invited to consult the recent book by J. Benesty, J. Chen, and Y. Huang, *Microphone Array Signal Processing*, Berlin, Germany: Springer-Verlag, 2008, and the references therein.

Several books on speech enhancement can be found in the literature. But they either do not cover state-of-the-art noise reduction techniques or are not rigorous in the explanations on how and why the noise reduction algorithms really work. This is how the idea of writing a book on *Noise Reduction in Speech Processing* came up. Therefore, the other objective of this text is to derive the most important and well-known techniques in this area of research and engineering in a rigorous way yet clear way, and prove many fundamental and intuitive results often taken for granted.

This book is especially written for graduate students and research engineers who work on noise reduction for speech and audio applications and want to understand the subtle mechanisms behind each approach. We hope the readers will find many new and interesting concepts that are presented in this book useful and inspiring.

VI   Preface

We would like to take this opportunity to thank again Christoph Baumann, Carmen Wolf, and Petra Jantzen from Springer (Germany) for their wonderful help in the preparation and publication of this manuscript. Working with them is always a pleasure and a wonderful experience. We also thank Dr. Emanuël Habets for reading a first draft of this book and giving us many useful comments. Finally, we deeply thank Huawei Technologies, Shenzhen, China, and especially the Media & Communication Lab of Huawei Research, for sharing with us some of their visions of the telecommunications of the future.

Montréal, QC/ Murray Hill, NJ/ Bridgewater, NJ/ Haifa   *Jacob Benesty*
Jingdong Chen
Yiteng Huang
Israel Cohen

Contents

1 Introduction ... 1
  1.1 Noise Reduction in Speech Processing ... 1
  1.2 The Paradigm for Noise Reduction ... 6
  1.3 A Brief History of Noise Reduction Research ... 7
  1.4 Organization of the Book ... 10
  1.5 Some Notes to the Reader ... 13

2 Problem Formulation ... 15
  2.1 In the Time Domain ... 15
  2.2 In the Frequency Domain ... 16
  2.3 In the Karhunen-Loève Expansion (KLE) Domain ... 18
  2.4 Summary ... 20

3 Performance Measures ... 21
  3.1 Signal-to-Noise Ratio ... 21
  3.2 Noise-Reduction Factor ... 24
  3.3 Speech-Distortion Index ... 25
  3.4 Speech-Reduction Factor ... 27
  3.5 Discussion ... 28

4 Mean-Squared Error Criterion ... 31
  4.1 In the Time Domain ... 31
  4.2 In the Frequency Domain ... 32
  4.3 In the KLE Domain ... 34
  4.4 Summary ... 36

5 Pearson Correlation Coefficient ... 37
  5.1 Correlation Coefficient Between Two Random Variables ... 37
  5.2 Correlation Coefficient Between Two Random Vectors ... 38
  5.3 Frequency-Domain Versions ... 39
  5.4 KLE-Domain Versions ... 39

VIII   Contents

5.5   Summary.. 40

6   **Fundamental Properties** 41
6.1   In the Time Domain 41
6.2   In the Frequency Domain.............................. 46
6.3   In the KLE Domain 50
6.4   Summary... 57

7   **Optimal Filters in the Time Domain**..................... 59
7.1   Wiener Filter 59
7.2   Tradeoff Filters 64
7.3   Subspace Approach 67
7.4   Experiments 68
7.4.1   Experimental Setup 68
7.4.2   Effect of Forgetting Factor on Performance 70
7.4.3   Effect of Filter Length on Performance 72
7.4.4   Performance in Different Noise Conditions 74
7.5   Summary... 75

8   **Optimal Filters in the Frequency Domain**................. 77
8.1   Wiener Filter 77
8.2   Parametric Wiener Filter 81
8.3   Tradeoff Filter 82
8.4   Experiments 86
8.4.1   Impact of Input SNR on Filter Gain and Speech
Distortion 86
8.4.2   Noise Estimation 86
8.4.3   Performance Comparison in NYSE Noise 89
8.4.4   Performance Comparison in Car Noise 93
8.5   Summary... 94

9   **Optimal Filters in the KLE Domain** 95
9.1   Class I 95
9.1.1   Wiener Filter 95
9.1.2   Parametric Wiener Filter 100
9.1.3   Tradeoff Filter 101
9.2   Class II 105
9.2.1   Wiener Filter 105
9.2.2   Tradeoff Filter 108
9.3   Experiments 111
9.3.1   Impact of Forgetting Factor on Performance of Class-I
Filters 111
9.3.2   Effect of Filter Length on Performance of Class-I
Filters 113
9.3.3   Estimation of Clean Speech Correlation Matrix 114

Contents IX

| | | 9.3.4 | Performance of Class-I Filters in Different Noise Conditions | .... | 116 |

9.3.4 Performance of Class-I Filters in Different Noise
 Conditions 116
9.3.5 Impact of Forgetting Factor on Performance of
 Class-II Filters 116
9.3.6 Effect of Filter Length on Performance of Class-II
 Filters 120
9.4 Summary... 121

10 Optimal Filters in the Transform Domain 123
10.1 Generalization of the KLE 123
10.2 Performance Measures 127
 10.2.1 SNR 127
 10.2.2 Noise-Reduction Factor 128
 10.2.3 Speech-Distortion Index 129
 10.2.4 Speech-Reduction Factor 130
10.3 MSE Criterion 131
10.4 PCC and Fundamental Properties 132
10.5 Examples of Filter Design 138
 10.5.1 Wiener Filter 138
 10.5.2 Parametric Wiener Filter 142
 10.5.3 Tradeoff Filter 143
 10.5.4 Examples of Unitary Matrices....................... 145
10.6 Experiments 146
 10.6.1 Performance of Wiener Filter in White Gaussian Noise.. 146
 10.6.2 Effect of Filter Length on Performance 148
 10.6.3 Performance of Tradeoff Filter in White Gaussian
 Noise 148
10.7 Summary... 152

11 Spectral Enhancement Methods 153
11.1 Problem Formulation 153
11.2 Performance Measures 155
 11.2.1 SNR 155
 11.2.2 Noise-Reduction Factor 156
 11.2.3 Speech-Distortion Index 157
 11.2.4 Speech-Reduction Factor 158
11.3 MSE Criterion 159
11.4 Signal Model... 161
11.5 Signal Estimation 162
 11.5.1 MMSE Spectral Estimation 162
 11.5.2 MMSE Spectral Amplitude Estimation 163
 11.5.3 MMSE Log-Spectral Amplitude Estimation 164
11.6 Spectral Variance Model 166
 11.6.1 GARCH Model 166
 11.6.2 Modeling Speech Spectral Variance 168
 11.6.3 Model Estimation 169

X   Contents

11.7 Spectral Variance Estimation 170
  11.7.1 Relation to Decision-Directed Estimation 172
11.8 Summary of Spectral Enhancement Algorithm 174
11.9 Experimental Results 176
11.10 Summary 181

12 A Practical Example: Multichannel Noise Reduction for Voice Communication in Spacesuits 183
12.1 Problem Description 183
12.2 Problem Analysis 186
  12.2.1 Sources of Noise in Spacesuits 186
  12.2.2 Noise Cancelling Microphones 188
12.3 Suggested Algorithms 192
  12.3.1 Nearfield, Wideband Microphone Array Beamforming for Speech Acquisition in Spacesuits 193
  12.3.2 Multichannel Noise Reduction: a More Practical Microphone Array Signal Processing Technique 200
  12.3.3 Single-Channel Noise Reduction 201
  12.3.4 Adaptive Noise Cancellation 202
12.4 Algorithm Validation 203
  12.4.1 In-Helmet Multichannel Acoustic Data Collection 203
  12.4.2 Performance Evaluation of Beamforming Algorithms ... 208
  12.4.3 Validation of Multichannel Noise Reduction Algorithms 214
  12.4.4 Validation of Single-Channel Noise Reduction Algorithms 214
  12.4.5 Feasibility Assessment of Using Adaptive Noise Cancellation in Spacesuits 214
12.5 Summary 217

References 219

Index 227

1

Introduction

1.1 Noise Reduction in Speech Processing

The world we live in is full of different kinds of sounds; it is obviously hard to imagine a world without any sound sources. Some sounds may originate from far-away sources but they die away before reaching our ears so we do not hear them. Some others, however, reach our ears, but their energies are so weak that we do not perceive them clearly. There are sounds that we can clearly hear. Among those, some may convey very important information while others may not carry anything really useful, but purely interfere with the sounds that do contain useful information. The sound that carries the information that we want or need, is normally labelled as "desired." The sound that does not contain any useful information but interferes with the desired sound, is usually referred to as "noise."

The term "noise" was coined in 1905 by Einstein when he saw that it could be the instrument to establish the existence of atoms [34]. Ever since, noise has permeated every field of science and technology and has served as a driving force for countless great discoveries and inventions, such as the existence of atoms, the Brownian motion, the Big Bang theory, the information theory, the optimal filtering theory, to name a few. There are many exciting stories behind noise [34]. One example of great interest is the Big Bang story. In the 1930s, Bell Laboratories built a large rotating antenna in Crawford Hill, New Jersey, to identify and understand various sources of noise that impaired the then-recently introduced transatlantic radio-telephone service. Having eliminated all known sources, the team found that there was a persistent residual "hiss" no matter where the antenna was pointed. At that time there were two competing theories of the universe. In one corner was the Big Bang model, which proposed a unique moment of creation followed by a rapid expansion. Hubble had observed that galaxies were moving away from each other and that the further they are, the faster they were moving away. Also, Gamow and Alpher had shown that the Big Bang could explain the abundance of hydrogen and helium. They and others predicted that the primordial radia-

J. Benesty et al., *Noise Reduction in Speech Processing*, Springer Topics in Signal Processing 2, DOI 10.1007/978-3-642-00296-0_1, © Springer-Verlag Berlin Heidelberg 2009

2   1 Introduction

tion, according to the Big Bang model, should be detectable as an isotropic field with a black body temperature of about 10 K. In the other corner was the Steady-State model, invented by Hoyle, Gold, and Bondi, which harked back to the conservative view of an eternal universe, except that it included an element of continuous creation and expansion. In the earlier 1960s, Wilson and Penzias, who were working at Bell Laboratories, detected the isotropic residual radiation using the 6-meter aperture antenna at Crawford Hill. When Penzias called Dicke (who was a prominent physicist at Princeton University and was working on how to find and measure the leftover radiation at that time) about what was measured and to seek advice on a possible explanation, he immediately realized what Penzias and his colleague Wilson had discovered. This example shows the importance of noise in the confirmation of the Big Bang theory.

However, as noise permeates every research and engineering field, its definition also evolves. Now, many fields have their own definition and history of noise, and in most cases, the definition has well departed from its original meaning as basically an idea, a subject, a field, or an instrument. Such an evolution, although accompanied with distortion, does make the description and understanding of the problems in a specific field easier. For the problems covered in this book, we follow the tradition built in the signal and speech processing fields and the term noise is used to signify any unwanted signal that interferes with measurement, processing, and communication of the desired information-bearing speech signal.

With this definition of noise, it is then of great interest to know the impact of noise on speech communications. As a matter of fact, this has been one of the most studied problems in human-to-human and human-to-machine communications over the last four decades. This problem, as perhaps we all know, is very much complicated because noise can cause profound changes to the characteristics of the speech signal. These changes can be classified into two categories, depending on in what stage they are introduced during a communication process.

- Changes in the talking/transmitting side. The talker, when hears the background noise, tends to alter his/her speaking style in an effort to increase communication efficiency over the noisy medium. This phenomenon is called the Lombard effect. This effect leads to changes in both speech intensity and characteristics. The intensity change has been experimentally quantified by, among others, Pearsons, Bennett, and Fidell [99]. Briefly, in many noisy environments, the speaker would increase his/her vocal intensity by 0.6 decibel (dB) for each dB increase in the background noise up to a ceiling level. However, the changes in speech characteristics are very difficult to model.
- Changes in the listening/receiving end. In the Lombard effect, noise causes changes in speech statistics through an indirect way. But in most situations, noise may directly affect the speech signal. This has two forms. One

1.1 Noise Reduction in Speech Processing 3

is when the noise and speech signals are statistically dependent. As a result, the additive noise directly modifies the spectrum and characteristics of the speech signal. The other is when the noise is independent of the speech signal, so it does not cause any changes to the speech characteristics. However, the characteristics of the observed signal is very different from those of the desired speech since the observed signal is a mixture of the desired speech and noise.

Regardless how the speech characteristics are modified by the presence of noise, the noise effect will profoundly affect the listener's perception and machine's processing of the observed speech. On the positive side, if the intensity of the noise is not too high, voice communication is still possible, but it would be less natural and less comfortable. On the negative side, if the noise is strong and the speech signal is completely immersed into it, voice communication would become difficult and even impossible. Therefore, in order to make voice communication feasible, natural, and comfortable in the presence of noise regardless of the noise level, it is desirable to develop digital signal processing techniques to "clean" the recorded signal before it is stored, transmitted, or played out. This problem is generally referred to as either noise reduction or speech enhancement.

The fact that noise can be any unwanted signal makes noise reduction an extremely difficult problem to tackle. As of today, it is still not clear how this problem can be described into a general mathematical model, not even mention the solutions. So, instead of searching for a generic solution, researchers and engineers in the speech field are adopting a pragmatic approach: dividing noise into several categories based on the mechanism how it is generated and conquering each category using a different approach. Commonly, noise is divided into four basic categories as explained below.

- *Additive noise*. Additive noise can come from various sources. Some are from natural sound sources, while others may be artificially introduced, for example, the comfort noise in speech codecs.
- *Echo*. Acoustic echo occurs due to the coupling between loudspeakers and microphones. The existence of echo will make conversations very difficult or even impossible.
- *Reverberation*. Reverberation is the result of multipath propagation and is introduced by an enclosure. It can cause spectral distortion, which impairs speech intelligibility.
- *Interference*. Interference comes from concurrent sound sources. In teleconferencing and telecollaboration, it is possible that each communication site has multiple participants and loudspeakers. So, there can be multiple competing sound sources.

Combating these four categories of noise has led to the developments of diverse acoustic signal processing techniques. They include noise reduction (or

4   1 Introduction

speech enhancement), echo cancellation and suppression, speech dereverberation, and source separation, each of which is a rich subject of research. A broad coverage of these research areas can be found in [15], [72]. This monograph will mainly focus on the (additive) noise reduction, which aims at eliminating or mitigating the effects of additive noise. So, from now on, we shall narrow the definition of noise down to additive noise.

Under this narrowed definition, the observed microphone signal can be modeled as a superposition of the clean speech and noise. The objective of noise reduction, then, becomes to restore the original clean speech from the mixed signal. This can be described as a parameter estimation problem and the optimal estimate of the clean speech can be achieved by optimizing some criterion, such as the mean-squared error (MSE) between the clean speech and its estimate, the signal-to-noise ratio (SNR), the *a posteriori* probability of the clean speech given its noisy observations, etc. Unfortunately, an optimal estimate formed from a signal processing perspective does not necessarily correspond to the best quality when perceived by the human ear. This inconsistency between objective measures and subjective quality judgement has forced researchers to rethink performance criteria for noise reduction. The objective of the problem has subsequently been broadened, which can be summarized as to achieve one or more of the following three primary goals:

- to improve objective performance criteria such as intelligibility, SNR, etc.,
- to improve the perceptual quality of a degraded speech, and
- as a preprocessor, to increase robustness of other speech processing applications (such as speech coding, echo cancellation, automatic speech recognition, etc.) to noise.

The different goals may lead to distinct speech estimates. It is very hard to satisfy all three goals at the same time. With a specified goal (performance criterion), the difficulty and complexity of the noise reduction problem can vary tremendously, depending on many factors such as the number of microphone channels. In general, the larger the number of microphones, the easier the noise reduction problem. For example, when an array of microphones can be used, a beam can be formed and steered to a desired direction. As a result, signal propagating from the desired direction will be passed through without degradation, while signals originating from all other directions will either suffer a certain amount of attenuation or be completely rejected [15]. In the two-microphone case, with one microphone picking up the noisy signal and the other measuring the noise field, we can use the second microphone signal as a noise reference and eliminate the noise in the first microphone by means of adaptive cancellation [6]. However, most of today's communication terminals are equipped with only one microphone. In this case, the noisy speech is the only resource that is accessible to us, so noise reduction becomes a very challenging problem for many sophisticated reasons. First, a reference of the noise is not accessible and the clean speech cannot be preprocessed prior to being corrupted by noise. Second, the nature and characteristics of the noise

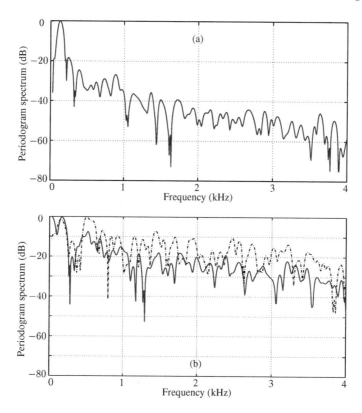

Fig. 1.1. Spectra of two different noise signals: (a) periodogram spectrum of a car noise signal and (b) periodogram spectra of noise in a conference room measured at two time instants (two seconds apart).

can change dramatically from application to application and even vary in time in one application. To illustrate this, Fig. 1.1 shows the spectra of a car noise signal and a noise signal recorded in a conference room [spectra computed at two different time instants (two seconds apart)]. It is seen that the spectrum of the car noise is very different from that of the conference room noise. Even for the same conference room, the characteristics of the noise are different when measured at different points in time. In addition, the speech signal is highly nonstationary. It is very difficult to estimate the fast changing statistics of speech in the presence of noise. Despite these challenges, a tremendous amount of attention has been dedicated to this problem since there are so many applications that require a solution to it.

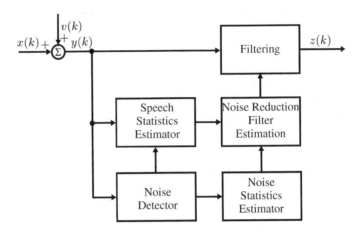

Fig. 1.2. General block diagram of a noise reduction system.

1.2 The Paradigm for Noise Reduction

This book addresses the challenging problem of how to achieve noise reduction by analyzing and processing the noisy speech measured by only a single microphone without requiring additional information. The general model used throughout this book is shown in Fig. 1.2. The model begins with a clean speech signal (from a desired speaker), $x(k)$, being corrupted with an unwanted additive noise, $v(k)$. The noisy signal, $y(k)$, which is a superposition of $x(k)$ and $v(k)$ is first processed to determine whether the desired speech is currently present or absent. The noise and speech statistics such as the covariance matrices and power spectral densities are estimated based on the detection results and the input noisy speech. These statistics will be used to estimate a noise reduction filter. This filter can be optimal in the sense that it optimizes some error criterion (e.g., MSE). It may be suboptimal, where parameters are introduced to better control the quality of the output speech. The estimated filter is applied to the noisy speech to filter out the noise signal, thereby producing an output signal, $z(k)$, which is supposed to be an estimate of the clean speech, $x(k)$.

The problem addressed in this book is of great importance from both the theoretical and practical viewpoints. To illustrate this, we give the very good example of multiparty conferencing explained by Diethorn in [38]. In multiparty conferencing, the background noise picked up by the microphone of each point of the conference combines additively at the network bridge with the noise signals from all other points. The loudspeaker at each location of the conference therefore reproduces the combined sum of the noise sequences from all other locations. Consider a three-point conference in which the room noise at all locations is stationary and independent with power σ_v^2. Each loudspeaker receives noise from the other two locations, resulting in a total noise power that is $2\sigma_v^2$, which is 3 dB greater than that of a two-point conference. Now if we consider a P-point conference, each side receives a total noise power of $(P-1)\sigma_v^2$, which is $10\log_{10}(P-1)$ dB greater than σ_v^2. This is still the ideal condition. In practice, there may be many processing operations in the voice communication network, such as speech coding, transcoding, automatic-gain-control processing, etc. Each operation may boost the noise level. So, the noise problem is extremely serious, particularly when the number of conferees is large, and without noise reduction, communication in this context is almost impossible.

Besides multiparty conferencing, there are many other applications that require a noise reduction technology. The following is a short list:

- hands-free communications,
- hearing aids,
- audio bridging,
- teleconferencing and telepresence,
- hands-free human-machine interfaces,
- car and mobile phones,
- cockpits and noisy manufacturing,
- high-quality speech coding.

To summarize this section, we can definitely claim that there are many speech related applications that require a technique to reduce noise. Greater efforts from both the research and engineering communities are indispensable to develop practical and reliable solutions before the noise reduction technology can be widely deployed.

1.3 A Brief History of Noise Reduction Research

Research on noise reduction has been going on for more than four decades. A significant amount of progress has been achieved over this period. To gain an appreciation of the basic concepts and fundamental techniques that were developed, it is worthwhile to briefly review some research highlights and milestones. The reader, however, is cautioned that such a review is narrowly focused and might not be comprehensive.

The earliest attempts to devise noise reduction algorithms for enhancing speech signals were made in the 1960s. In 1960, at Bell Laboratories, Schroeder proposed a system for reducing noise in telecommunication environments. The schematic diagram of his system is shown in Fig. 1.3. Note that this diagram is slightly modified from its original form in [108] for ease of exposition. As seen, the input noisy signal, $y(t)$ (where t is the continuous time), which is a superposition of the clean speech, $x(t)$, and noise, $v(t)$, i.e., $y(t) = x(t)+v(t)$, is divided into M subbands. For each subband, a rectifier and a lowpass filter are applied in tandem to estimate the noisy speech envelope. The noise level in the

1 Introduction

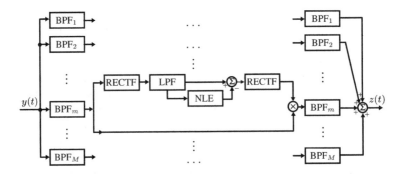

Fig. 1.3. Schematic diagram of the Schroeder's noise reduction system where BFP, RECTF, LFP, and NLE denote, respectively, bandpass filter, rectifier, lowpass filter, and noise level estimator.

corresponding subband is then estimated using an analog circuit with resistors, capacitors, and diodes, and the noise estimate is subsequently subtracted from the noisy speech envelope, resulting in an estimate of the clean speech envelope for the subband. A second rectification process is applied to force the negative results to zero due to the subtraction. The rectified clean speech envelope estimate, which is served as a gain filter, is then multiplied with the unmodified subband signal. Finally, the fullband signal, $z(t)$, is constructed from all the subband outputs, where $z(t)$ is basically an estimate of $x(t)$. As seen, the Schroeder's system is actually a spectral subtraction technique, but with all analog implementation. However, this work was largely unknown in the noise reduction research community until more formal methods were developed in the late 1970s.

The 1970s was a decade in which great impetus was given to the spectral modification based techniques. This wave of interest in noise reduction was due largely to the confluence of digital signal processing (DSP) algorithms and DSP hardwares, both of which had come to prominence at that time. In 1974, Weiss, Aschkenasy, and Parsons developed a "spectrum shaping" method that used amplitude clipping in filter banks to remove low-level excitation, presumably noise [122]. A few years later, Boll, in his informative paper [18], reinvented the spectral subtraction method but in the digital domain. Boll was perhaps the first to formulate the magnitude subtraction in the framework of digital short-time Fourier analysis, which had earlier been under development by, among others, Allen [1] and Portnoff [101]. Shortly after, McAulay and Malpass cast the spectral subtraction approach in a framework of statistical spectral estimation, and presented a broad class of estimators including the magnitude and power subtraction, the Wiener filtering, the maximum likelihood envelope estimator, etc. [92]. They were also the first to make connections between the spectral subtraction and the Wiener filter. Almost at the same time, Lim and Oppenheim, in their landmark work [85], presented

1.3 A Brief History of Noise Reduction Research 9

one of the first comprehensive treatments of methods of noise reduction and speech enhancement. The spectrum subtraction methods were discussed, also within an estimation framework, and were compared to other techniques of speech enhancement. Also, Sondhi, Schmidt, and Rabiner published results from a series of implementation studies that roots in Schroeder's work in the 1960s [112].

In the 1980s, many fundamental ideas in noise reduction surfaced and were published. By and large, these ideas can be broadly classified into two categories, namely, spectral restoration based on estimation theory and speech model based methods. The spectral restoration technique treats noise reduction as a robust spectral estimation problem, i.e., estimating the spectrum of the clean speech from that of the noisy speech. In 1984, Ephraim and Malah developed an optimal spectral amplitude and some spectral phase estimators using the statistical estimation theory [44]. Their paper was widely cited in the noise reduction research and (together with McAulay and Malpass' work) was most notable as an early attempt to apply statistical theory to noise reduction. It is also the first showing, by theory, that the optimal estimate of the clean speech spectral phase is the spectral phase of the noisy speech. Therefore, the basic problem of noise reduction becomes one of estimating only the clean speech spectral amplitude (this result was used previously but mainly based on experimental observations). Following this work, many statistical spectral estimators were developed, including the minimum mean-squared error (MMSE) log-spectral amplitude estimator, the maximum-likelihood (ML) spectral amplitude estimator, the ML spectral power estimator, the maximum a posteriori (MAP) spectral amplitude estimator, etc. Today, there are still tremendous efforts to search for better spectral amplitude estimators, which are inspired by McAulay and Malpass' and Ephraim and Malah's work. The model-based approaches also formulate noise reduction as an estimation problem. Comparatively, in this category of methods, a mathematical model is used to represent human speech production and parameter estimation is carried out in the model space, which normally has a much lower dimensionality than the original signal space. Lim and Oppenheim studied the harmonic (or sinusoidal) model for speech and developed a method to reduce noise using comb filtering [86]. In 1987, Paliwal and Basu published a paper that used the linear predictive coding (LPC) model and the Kalman filter to reduce noise [97]. The basic idea underlying the LPC model is that a given speech sample at time k can be approximated as a linear combination of the past few speech samples. The LPC model was well studied and widely used in speech analysis, speech coding, and speech recognition in the 1980s, but Paliwal and Basu were the first to combine the LPC model with the Kalman filter for noise reduction. Since then, many efforts have been made to make this technique more practical.

A number of significant milestones were achieved in the 1990s, from both the theoretical and application points of view. From the theoretical side, the hidden Markov models (HMMs) were borrowed from speech recognition and

10   1 Introduction

applied to noise reduction [46], [47], [48]. The HMM based method is basically similar to the statistical spectral estimator developed in the 1980s. The difference is that the statistical spectral estimators assume the explicit knowledge of the joint probability distribution of the clean speech and noise signals so that the conditional expected value of the clean speech (or its sample spectrum), given the noisy speech, can be evaluated, while in the HMM based approach, the unknown probability distributions of the speech and noise are learned from the training sequence. Also, subspace methods were developed by Dendrinos, Bakamidis, and Carayannis [37] and by Ephraim and Van Trees [49]. In essence, the subspace approach projects the noisy signal vector into a different domain either via the Karhunen-Loève (KL) transform through eigenvalue decomposition of an estimate of the correlation matrix of the noisy signal [49] or by using the singular value decomposition of a Toeplitz-structured data matrix specially arranged from the noisy signal vector [78]. Once transformed, the speech signal only spans a portion of the entire space, and as a result, the entire vector space can be divided into two subspaces: the signal-plus-noise and the noise only. The noise statistics can then be estimated from the noise only subspace. From the application side, the noise reduction technology has been increasingly used in telecommunications. Many parametric codecs, such as the enhanced variable rate codec (EVRC), the adaptive multi-rate (AMR) codec, etc., integrated noise reduction into speech compression.

1.4 Organization of the Book

The material in this book is organized into twelve chapters, including this one. We attempted to cover the most basic concepts and fundamental techniques used in noise reduction from a signal processing perspective in the next eleven chapters. The material discussed in these chapters is as follows.

Chapter 2 provides an overview of the noise reduction problem. The most straightforward and classical approach to tackling this problem is through linear filtering in the time (original signal) domain. However, it is often very advantageous to deal with this problem in a transform space. This is due to the fact that if the transform is properly selected, the speech and noise signals can be better separated in that space, making it easier to estimate the noise statistics, and thereby optimizing the noise reduction filter. So, we discuss how to model the problem not only in the time domain, but also in the two well-known transform spaces, i.e., the frequency and Karhunen-Loève expansion (KLE) domains.

A key issue in the design and implementation of noise reduction filters is how to assess their performance. Chapter 3 focuses on the performance metrics. We present several useful measures based on signal processing techniques. These measures will not only help properly design filters in the different domains, but also help us understand how noise reduction works in real-world applications and what price we need to pay for achieving noise reduction.

1.4 Organization of the Book   11

Most noise reduction filters are derived by optimizing some error criteria. The most used error criterion is, by far, the MSE. In Chapter 4, we present the MSE and its normalized form in the time, frequency, and KLE domains. We will demonstrate that the normalized form, called normalized MSE (NMSE, which can be in either subband or fullband) in all domains depends explicitly on the input SNR, and some other performance measures presented in Chapter 3, which makes the NMSE not only useful in deriving different optimal filters, but also powerful in analyzing the noise reduction performance.

In the context of noise reduction, it is very important to directly compare the output SNR with the input SNR. This is essential in order to tell whether the filter is indeed able to reduce noise relative to the speech level. However, the MSE and NMSE, which are overwhelmingly used to derive noise reduction filters, are not explicitly related to the output SNR even though intuitively they should depend on it. In Chapter 5, we present an alternative criterion to the MSE, called the squared Pearson correlation coefficient (SPCC), in which the output SNR appears naturally. We develop several forms of this coefficient in the different domains.

In Chapter 6, we discuss many interesting properties of the SPCC in the different domains. These fundamental properties establish relationships between the SPCC, the input SNR, the output SNR, and several other performance measures. Armed with these properties it is possible in many cases to derive and analyze optimal noise reduction filters by a simple inspection of a certain form of the SPCC.

In Chapter 7, we derive various forms of noise reduction filters in the time domain. The major focus is on the Wiener filter, which is an optimal filter in the minimum MSE (MMSE) sense, and many known algorithms are related to this filter in one way or another. We discuss many fundamental and interesting properties of this filter and demonstrate that the Wiener filter achieves noise reduction at the cost of adding speech distortion. To manage the compromise between noise reduction and speech distortion, we present a tradeoff filter, which is derived from a constrained MSE criterion where a parameter is used to adjust the compromise between the amount of noise reduction and the amount of speech distortion. We also discuss the well-known subspace method, which can be viewed as a special way of implementing the tradeoff filter.

While the time domain is the most straightforward domain to work with, the derived noise reduction filters are often less flexible in terms of performance tuning. For example, in practice, noise is not necessarily white and in many cases its energy may concentrate only in some frequencies or frequency bands. In such situations, it is advantageous to design noise reduction filters on a subband basis. In Chapter 8, we investigate the design of noise reduction filters in the frequency domain, which is an alternative to the time domain. The frequency-domain filters are, so far, the most popular approaches to noise reduction because: 1) the filters at different frequencies (or frequency bands) are designed and used independently with each other, this design offers sig-

12   1 Introduction

nificant flexibility in dealing with colored and band-limited noise; 2) most of
our knowledge and understanding of speech production and perception is related to frequencies; and 3) thanks to the fast Fourier transform (FFT), the
implementation of frequency-domain filters are generally very efficient. We
develop some widely used classical frequency-domain filters and discuss their
properties.

Although it is widely studied, the frequency domain is not the only domain
we can work with. There are other domains that may offer more advantages
for the design of noise reduction algorithms. In Chapter 9, we consider the
KLE domain. We derive two broad classes (the classification is basically based
on the subband filter length) of optimal filters in this domain. The first class,
similar to the frequency-domain filters, estimates a frame of the clean speech
by filtering the corresponding frame of the noisy speech while the second
class does noise reduction by filtering not only the current frame but also a
number of previous consecutive frames of the noisy speech. We demonstrate
that better noise reduction performance can be achieved with the Class II
filters when the parameters associated with this class are properly chosen.

Chapter 10 is basically an extension of Chapter 9. In this chapter, we introduce a new transform domain, where any unitary (or orthogonal) matrix
can be used to construct the forward (for analysis) and inverse (for synthesis)
transforms. This new domain can be viewed as a generalization of the KLE
domain. The advantages of working in this generalized domain are multiple,
such as different transforms can be used to replace each other without any requirement to change the algorithm formulation (optimal filter) and it is easier
to fairly compare different transforms for their noise reduction performance.
We address the design of different optimal and suboptimal filters in such
a generalized transform domain, including the Wiener filter, the parametric
Wiener filter, some tradeoff filters, etc. We also compare different transforms,
including the KL, Fourier, cosine, and Hadamard transforms, for their noise
reduction performance.

Chapter 11 can be seen as an extension of Chapter 8. Chapter 8 considers
the optimal noise reduction filters in the continuous frequency domain where
the frequency-domain representation of a discrete speech signal is obtained
through the discrete-time Fourier transform (DTFT). In Chapter 11, we shift
our focus from the DTFT domain to the discrete Fourier transform (DFT)
domain, where the frequency-domain representation of the input speech signal
is achieved through the short-time discrete Fourier transform (STDFT), or
simply the short-time Fourier transform (STFT). Using the STFT as opposed
to DTFT, we have to take many effects such as circular convolution (which
may cause the aliasing problem) and finite frame length (which may change
the statistical distribution of the spectrum) into consideration. We formulate
the noise reduction problem in the STFT domain as both a filtering and
estimation problem. We present statistical models for the speech and noise
signals in this domain and derive estimators for the speech signal using various
distortion measures.

Noise reduction can find numerous applications in various fields of speech processing. It is, however, a very difficult problem since its technical objectives and the acoustic environments can vary significantly from one application to another. As a result, it is very difficult, if not impossible, to find a versatile algorithm that can really work for all the applications in different practical environments. To illustrate the difficulty of this problem, we provide an example in Chapter 12, where we consider the noise reduction problem in the helmet of an astronaut's spacesuit. Of course, the major focus of this chapter is not the finding of the best noise reduction algorithm for the specific problem; instead, we attempt to provide some deep analysis of the acoustic challenges and the validation procedure of the suggested solutions that can be more useful and can enrich the reader's fund of knowledge. Although we restrict our attention in this book to noise reduction using a single microphone, we will also briefly examine the possibility of using multiple microphones to achieve noise reduction in this chapter, while most of the studied multichannel techniques can be found in the book *Microphone Array Signal Processing* [15].

1.5 Some Notes to the Reader

Successful noise reduction systems require the knowledge and expertise from a wide range of disciplines such as signal processing, statistics, estimation and detection theory, pattern recognition, speech processing, communications, and speech perception, to name a few. It is very difficult for any single person to master all these disciplines. The philosophy and purpose of this book is to provide in-depth discussions of a number of fundamental topics in noise reduction research from a signal processing viewpoint so that the reader can have a good understanding of the fundamentals of this problem without necessarily having to be an expert in all disciplines enumerated above. Our basic goal is to obtain optimal filters for noise reduction and we resort to suboptimal filters if the compromise between the amount of noise reduction and the amount of speech distortion has to be made. We have tried to give deep insights into the noise reduction problem by including results that we deem to be most useful and important in practice.

The mathematical notation for all symbols is defined within the context of the discussion and when it first appears in the text. In general, (column) vectors are bold lowercase and matrices are bold uppercase.

2

Problem Formulation

In this chapter, we explain the traditional formulation of the (additive) noise problem and its reduction by linear filtering in the time, frequency, and Karhunen-Loève expansion domains.

2.1 In the Time Domain

The noise reduction problem consists of recovering the signal of interest (clean speech or desired signal) $x(k)$ of zero mean from the noisy observation (microphone signal)

$$y(k) = x(k) + v(k), \tag{2.1}$$

where k is the discrete time index, and $v(k)$ is the unwanted additive noise, which is assumed to be a zero-mean random process (white or colored) and uncorrelated with $x(k)$.

The signal model given in (2.1) can be written in a vector form if we process the data by blocks of L samples:

$$\mathbf{y}(k) = \mathbf{x}(k) + \mathbf{v}(k), \tag{2.2}$$

where

$$\mathbf{y}(k) = \begin{bmatrix} y(k) \; y(k-1) \; \cdots \; y(k-L+1) \end{bmatrix}^T, \tag{2.3a}$$

$$\mathbf{x}(k) = \begin{bmatrix} x(k) \; x(k-1) \; \cdots \; x(k-L+1) \end{bmatrix}^T, \tag{2.3b}$$

$$\mathbf{v}(k) = \begin{bmatrix} v(k) \; v(k-1) \; \cdots \; v(k-L+1) \end{bmatrix}^T, \tag{2.3c}$$

and superscript ^{T} denotes transpose of a vector or a matrix. Since $x(k)$ and $v(k)$ are uncorrelated, the correlation matrix of the noisy signal is equal to the sum of the correlation matrices of the desired and noise signals, i.e.,

$$\mathbf{R}_y = \mathbf{R}_x + \mathbf{R}_v, \tag{2.4}$$

J. Benesty et al., *Noise Reduction in Speech Processing*, Springer Topics in Signal Processing 2,
DOI 10.1007/978-3-642-00296-0_2, © Springer-Verlag Berlin Heidelberg 2009

16   2 Problem Formulation

where

$$\mathbf{R}_y = E\left[\mathbf{y}(k)\mathbf{y}^T(k)\right], \tag{2.5a}$$
$$\mathbf{R}_x = E\left[\mathbf{x}(k)\mathbf{x}^T(k)\right], \tag{2.5b}$$
$$\mathbf{R}_v = E\left[\mathbf{v}(k)\mathbf{v}^T(k)\right], \tag{2.5c}$$

are the correlation matrices[1] of the signals $y(k)$, $x(k)$, and $v(k)$, respectively, and $E[\cdot]$ denotes mathematical expectation.

The objective of noise reduction is to estimate $\mathbf{x}(k)$ from the observation vector $\mathbf{y}(k)$. Usually, we estimate the noise-free speech, $\mathbf{x}(k)$, by applying a linear transformation to the microphone signal [10], [11], [25], [72], [88], [119], i.e.,

$$\mathbf{z}(k) = \mathbf{H}\mathbf{y}(k) \tag{2.6}$$
$$= \mathbf{H}\left[\mathbf{x}(k) + \mathbf{v}(k)\right]$$
$$= \mathbf{x}_\mathrm{F}(k) + \mathbf{v}_\mathrm{F}(k),$$

where $\mathbf{H}$ is a filtering matrix of size $L \times L$, $\mathbf{x}_\mathrm{F}(k) = \mathbf{H}\mathbf{x}(k)$ is the filtered clean speech (or filtered desired signal), and $\mathbf{v}_\mathrm{F}(k) = \mathbf{H}\mathbf{v}(k)$ is the filtered noise, which is often called the residual noise. The correlation matrix of the estimated signal is

$$\mathbf{R}_z = E\left[\mathbf{z}(k)\mathbf{z}^T(k)\right]$$
$$= \mathbf{H}\mathbf{R}_x\mathbf{H}^T + \mathbf{H}\mathbf{R}_v\mathbf{H}^T. \tag{2.7}$$

Therefore, with this time-domain formulation, the noise reduction problem becomes one of finding an optimal filter that would attenuate the noise as much as possible while keeping the clean speech from being dramatically distorted.

2.2 In the Frequency Domain

In the frequency domain, (2.1) can be rewritten as

$$Y(j\omega) = X(j\omega) + V(j\omega), \tag{2.8}$$

where j is the imaginary unit ($j^2 = -1$), and $Y(j\omega)$, $X(j\omega)$, and $V(j\omega)$ are respectively the discrete-time Fourier transforms (DTFTs) of $y(k)$, $x(k)$, and $v(k)$, at angular frequency ω ($-\pi < \omega \leq \pi$). Another possible form for (2.8) is

$$Y(\omega)e^{j\varphi_y(\omega)} = X(\omega)e^{j\varphi_x(\omega)} + V(\omega)e^{j\varphi_v(\omega)}, \tag{2.9}$$

---

[1] Here, we implicitly assume that all involved signals are stationary or quasi-stationary.

2.2 In the Frequency Domain   17

where for any random signal $A(j\omega) = A(\omega)e^{j\varphi_a(\omega)}$, $A(\omega)$ and $\varphi_a(\omega)$ are its amplitude and phase at frequency ω, $A \in \{Y, X, V\}$, $a \in \{y, x, v\}$. We recall that the DTFT and the inverse transform [96] are[2]

$$A(j\omega) = \sum_{k=-\infty}^{\infty} a(k)e^{-j\omega k}, \tag{2.10}$$

$$a(k) = \frac{1}{2\pi} \int_{-\pi}^{\pi} A(j\omega)e^{j\omega k} d\omega. \tag{2.11}$$

Using the power spectral density (PSD) of the noisy signal and the fact that $x(k)$ and $v(k)$ are uncorrelated, we get

$$\phi_y(\omega) = \phi_x(\omega) + \phi_v(\omega), \tag{2.12}$$

where

$$\phi_a(\omega) = E\left[|A(j\omega)|^2\right]$$
$$= E\left[A^2(\omega)\right]$$

is the PSD of the signal $a(k)$ [which is the inverse DTFT of $A(j\omega)$] at frequency ω.

An estimate of $X(j\omega)$ can be obtained by multiplying $Y(j\omega)$ with a complex gain, i.e.,

$$Z(j\omega) = H(j\omega)Y(j\omega) \tag{2.13}$$
$$= H(j\omega)\left[X(j\omega) + V(j\omega)\right]$$
$$= X_F(j\omega) + V_F(j\omega),$$

where $Z(j\omega)$ is the frequency-domain representation of the signal $z(k)$, and $X_F(j\omega) = H(j\omega)X(j\omega)$ and $V_F(j\omega) = H(j\omega)V(j\omega)$ are, respectively, the filtered clean speech and noise in the frequency domain. The PSD of $z(k)$ can then be written as

$$\phi_z(\omega) = |H(j\omega)|^2 \phi_y(\omega)$$
$$= |H(j\omega)|^2 \left[\phi_x(\omega) + \phi_v(\omega)\right]. \tag{2.14}$$

We can go back to the time domain using (2.11) to obtain the estimate

$$z(k) = \frac{1}{2\pi} \int_{-\pi}^{\pi} Z(j\omega)e^{j\omega k} d\omega. \tag{2.15}$$

---

[2] For the DTFT to exist, the time series $a(k)$ must be absolutely summable, i.e., $\sum_{k=-\infty}^{+\infty} |a(k)| < \infty$. This condition does not hold for stationary signals. Nevertheless, we will still use this spectral representation for stationary signals in the rest of this book for convenience of presentation and notation, knowing that the discrete time Fourier (DFT) will lead to similar theoretical results.

18   2 Problem Formulation

The objective of noise reduction in the frequency domain is then to find an optimal filter[3] $H(j\omega)$ at each frequency ω that would attenuate the noise greatly with as little distortion as possible to the desired signal (speech).

2.3 In the Karhunen-Loève Expansion (KLE) Domain

We first briefly recall the basic principle of the so-called Karhunen-Loève expansion (KLE) and then show how we can work in the KLE domain.

Let the $L \times 1$ vector $\mathbf{x}(k)$ denote a data sequence drawn from a zero-mean stationary process with the correlation matrix $\mathbf{R}_x$. This matrix can be diagonalized as follows [59]

$$\mathbf{Q}^T \mathbf{R}_x \mathbf{Q} = \mathbf{\Lambda}, \tag{2.16}$$

where

$$\mathbf{Q} = \begin{bmatrix} \mathbf{q}_1 & \mathbf{q}_2 & \cdots & \mathbf{q}_L \end{bmatrix}$$

and

$$\mathbf{\Lambda} = \text{diag} \begin{bmatrix} \lambda_1 & \lambda_2 & \cdots & \lambda_L \end{bmatrix}$$

are, respectively, orthogonal and diagonal matrices. The orthonormal vectors $\mathbf{q}_1, \mathbf{q}_2, \ldots, \mathbf{q}_L$ are the eigenvectors corresponding, respectively, to the eigenvalues $\lambda_1, \lambda_2, \ldots, \lambda_L$ of the matrix $\mathbf{R}_x$.

The vector $\mathbf{x}(k)$ can be written as a combination (expansion) of the eigenvectors of the correlation matrix $\mathbf{R}_x$ as follows

$$\mathbf{x}(k) = \sum_{l=1}^{L} c_{x,l}(k) \mathbf{q}_l, \tag{2.17}$$

where

$$c_{x,l}(k) = \mathbf{q}_l^T \mathbf{x}(k), \; l = 1, 2, \ldots, L \tag{2.18}$$

are the coefficients of the expansion and l is the subband index.

The representation of the random vector $\mathbf{x}(k)$ described by (2.17) and (2.18) is the Karhunen-Loève expansion (KLE) [63]. Equations (2.17) and (2.18) are, respectively, the synthesis and analysis parts of this expansion.

From (2.18), we can easily verify that

$$E\left[c_{x,l}(k)\right] = 0, \; l = 1, 2, \ldots, L \tag{2.19}$$

---

[3] As will become clearer in Chapter 8, such an optimal filter is in general real and positive, and its value is usually in the range between 0 and 1. So, when applied to the noisy speech, this filter only modifies the gain, but does not change the phase of $Y(j\omega)$. As a result, it is often called a gain filter.

2.3 In the Karhunen-Loève Expansion (KLE) Domain   19

and

$$E\left[c_{x,i}(k)c_{x,j}(k)\right] = \begin{cases} \lambda_i, & i = j \\ 0, & i \neq j \end{cases}. \tag{2.20}$$

It can also be checked from (2.18) that

$$\sum_{l=1}^{L} c_{x,l}^2(k) = \|\mathbf{x}(k)\|_2^2, \tag{2.21}$$

where $\|\mathbf{x}(k)\|_2$ is the Euclidean norm of $\mathbf{x}(k)$. The previous expression shows the energy conservation through the KLE process.

One of the most important aspects of the KLE is its potential to reduce the dimensionality of the vector $\mathbf{x}(k)$ for low-rank signals. This idea has been extensively exploited in different manners for noise reduction where the signal of interest (speech) is assumed to be a low-rank signal. In the following, we show how to work directly in the KLE domain.

Left-multiplying (2.2) by $\mathbf{q}_l^T$, we get

$$\begin{aligned} c_{y,l}(k) &= \mathbf{q}_l^T \mathbf{y}(k) \\ &= \mathbf{q}_l^T \mathbf{x}(k) + \mathbf{q}_l^T \mathbf{v}(k) \\ &= c_{x,l}(k) + c_{v,l}(k), \ l = 1, 2, \ldots, L. \end{aligned} \tag{2.22}$$

Expression (2.22) is equivalent to (2.2) but in the KLE domain. Again, we see that

$$\sum_{l=1}^{L} c_{y,l}^2(k) = \|\mathbf{y}(k)\|_2^2, \tag{2.23}$$

$$\sum_{l=1}^{L} c_{v,l}^2(k) = \|\mathbf{v}(k)\|_2^2. \tag{2.24}$$

We also have

$$E\left[c_{y,i}(k)c_{y,j}(k)\right] = \begin{cases} \lambda_i + \mathbf{q}_i^T \mathbf{R}_v \mathbf{q}_i, & i = j \\ \mathbf{q}_i^T \mathbf{R}_v \mathbf{q}_j, & i \neq j \end{cases}. \tag{2.25}$$

In the rest of this work, we assume that $|\mathbf{q}_i^T \mathbf{R}_v \mathbf{q}_j| \ll \lambda_i + \mathbf{q}_i^T \mathbf{R}_v \mathbf{q}_i$ or $|\mathbf{q}_i^T \mathbf{R}_v \mathbf{q}_j| \approx 0$, for $i \neq j$[4], so that we can estimate each one of the coefficients $c_{x,l}$, $l = 1, 2, \ldots, L$, without being related to the others. Clearly, our problem this time is to find an estimate of $c_{x,l}(k)$ by passing $c_{y,l}(k)$ through a linear filter, i.e.,

$$\begin{aligned} c_{z,l}(k) &= \mathbf{h}_l^T \mathbf{c}_{y,l}(k) \\ &= \mathbf{h}_l^T \left[\mathbf{c}_{x,l}(k) + \mathbf{c}_{v,l}(k)\right], \ l = 1, 2, \ldots, L, \end{aligned} \tag{2.26}$$

---

[4] If the noise is white, then $\mathbf{q}_i^T \mathbf{R}_v \mathbf{q}_j = 0$, $\forall i \neq j$.

20   2 Problem Formulation

where

$$\mathbf{h}_l = \begin{bmatrix} h_{l,0} \ h_{l,1} \ \cdots \ h_{l,L_l-1} \end{bmatrix}^T \tag{2.27}$$

is a finite-impulse-response (FIR) filter of length L_l, and

$$\mathbf{c}_{y,l}(k) = \begin{bmatrix} c_{y,l}(k) \ c_{y,l}(k-1) \ \cdots \ c_{y,l}(k-L_l+1) \end{bmatrix}^T, \tag{2.28a}$$

$$\mathbf{c}_{x,l}(k) = \begin{bmatrix} c_{x,l}(k) \ c_{x,l}(k-1) \ \cdots \ c_{x,l}(k-L_l+1) \end{bmatrix}^T, \tag{2.28b}$$

$$\mathbf{c}_{v,l}(k) = \begin{bmatrix} c_{v,l}(k) \ c_{v,l}(k-1) \ \cdots \ c_{v,l}(k-L_l+1) \end{bmatrix}^T. \tag{2.28c}$$

We see that the filters $\mathbf{h}_l$, $l = 1, 2, \ldots, L$, can take different lengths in the different subbands. The variance of the signal $c_{z,l}(k)$ is

$$E\left[c_{z,l}^2(k)\right] = \mathbf{h}_l^T \mathbf{R}_{c_x,l} \mathbf{h}_l + \mathbf{h}_l^T \mathbf{R}_{c_v,l} \mathbf{h}_l, \ l = 1, 2, \ldots, L, \tag{2.29}$$

where

$$\mathbf{R}_{c_x,l} = E\left[\mathbf{c}_{x,l}(k)\mathbf{c}_{x,l}^T(k)\right],$$

$$\mathbf{R}_{c_v,l} = E\left[\mathbf{c}_{v,l}(k)\mathbf{c}_{v,l}^T(k)\right],$$

are the correlation matrices of the signals $c_{x,l}(k)$ and $c_{v,l}(k)$, respectively. Apparently, reducing the noise in the KLE domain comes to the design of the FIR filters $\mathbf{h}_l$, $l = 1, 2, \ldots, L$. Finally, an estimate of the vector $\mathbf{x}(k)$ would be

$$\mathbf{z}(k) = \sum_{l=1}^{L} c_{z,l}(k)\mathbf{q}_l. \tag{2.30}$$

2.4 Summary

In this chapter, we have described the fundamental formulation of the noise reduction problem in the time, frequency, and KLE domains. This formulation will be frequently used throughout the rest of this book to derive the most well-known and useful noise reduction filters.

3

Performance Measures

In this chapter, we present some very useful measures that are necessary to properly design filters in the different domains. These definitions will also help us better understand how noise reduction works in real-world applications and what price we need to pay for this. As a matter of fact, all known approaches in the single-channel (one microphone) case add distortion to the desired signal. Therefore, in most applications, it is essential to find a good compromise between the amount of noise reduction and the degree of speech distortion, and the measures presented in this chapter can guide us to achieve such a compromise.

3.1 Signal-to-Noise Ratio

One of the most important measures in noise reduction is the signal-to-noise ratio (SNR). From the signal model given in (2.1), we define the input SNR as the ratio of the intensity of the signal of interest (speech) over the intensity of the background noise, i.e.,

$$\text{iSNR} = \frac{\sigma_x^2}{\sigma_v^2},\tag{3.1}$$

where

$$\sigma_x^2 = E\left[x^2(k)\right]$$

and

$$\sigma_v^2 = E\left[v^2(k)\right]$$

are the variances of the signals $x(k)$ and $v(k)$, respectively. This definition of the input SNR can also be written in different forms. With the signal model shown in (2.2), it is easy to check that

J. Benesty et al., *Noise Reduction in Speech Processing*, Springer Topics in Signal Processing 2,
DOI 10.1007/978-3-642-00296-0_3, © Springer-Verlag Berlin Heidelberg 2009

22   3 Performance Measures

$$\sigma_x^2 = \mathrm{tr}\left(\mathbf{R}_x\right)/L$$

and

$$\sigma_v^2 = \mathrm{tr}\left(\mathbf{R}_v\right)/L,$$

where $\mathrm{tr}(\cdot)$ denotes the trace of a matrix. Therefore, the input SNR can be written as

$$\mathrm{iSNR} = \frac{\mathrm{tr}\left(\mathbf{R}_x\right)}{\mathrm{tr}\left(\mathbf{R}_v\right)}. \tag{3.2}$$

According to the Parseval's relation, we have

$$\sigma_x^2 = \frac{1}{2\pi} \int_{-\pi}^{\pi} \phi_x(\omega)d\omega \tag{3.3}$$

and

$$\sigma_v^2 = \frac{1}{2\pi} \int_{-\pi}^{\pi} \phi_v(\omega)d\omega. \tag{3.4}$$

Substituting (3.3) and (3.4) into (3.1), we get another form of the input SNR:

$$\mathrm{iSNR} = \frac{\int_{-\pi}^{\pi} \phi_x(\omega)d\omega}{\int_{-\pi}^{\pi} \phi_v(\omega)d\omega}. \tag{3.5}$$

In the KLE domain, if we apply the matrix diagonalization of (2.16) to (3.2), we can readily deduce that

$$\mathrm{iSNR} = \frac{\sum_{l=1}^{L} \lambda_l}{\sum_{l=1}^{L} \mathbf{q}_l^T \mathbf{R}_v \mathbf{q}_l}. \tag{3.6}$$

In the frequency and KLE domains, it is also important to examine the SNR in each subband. So, we define the subband input SNRs in these two domains as

$$\mathrm{iSNR}(\omega) = \frac{\phi_x(\omega)}{\phi_v(\omega)}, \ \omega \in (-\pi, \pi] \tag{3.7}$$

and

$$\mathrm{iSNR}_l = \frac{\lambda_l}{\mathbf{q}_l^T \mathbf{R}_v \mathbf{q}_l}, \ l = 1, 2, \ldots, L. \tag{3.8}$$

After noise reduction with the time-domain model given in (2.6), the output SNR can be written as

$$\text{oSNR}(\mathbf{H}) = \frac{E\left[\mathbf{x}_\text{F}^T(k)\mathbf{x}_\text{F}(k)\right]}{E\left[\mathbf{v}_\text{F}^T(k)\mathbf{v}_\text{F}(k)\right]}$$

$$= \frac{\text{tr}\left(\mathbf{H}\mathbf{R}_x\mathbf{H}^T\right)}{\text{tr}\left(\mathbf{H}\mathbf{R}_v\mathbf{H}^T\right)}. \tag{3.9}$$

One of the most important goals of noise reduction is to improve the SNR after filtering [11], [24]. Therefore, we must design a filter, $\mathbf{H}$, in such a way that $\text{oSNR}(\mathbf{H}) \geq \text{iSNR}$.

In the frequency and KLE domains, the subband output SNRs are

$$\text{oSNR}\left[H(j\omega)\right] = \frac{|H(j\omega)|^2 \phi_x(\omega)}{|H(j\omega)|^2 \phi_v(\omega)}$$

$$= \text{iSNR}(\omega), \ \omega \in (-\pi, \pi] \tag{3.10}$$

and

$$\text{oSNR}(\mathbf{h}_l) = \frac{\mathbf{h}_l^T \mathbf{R}_{c_x,l}\mathbf{h}_l}{\mathbf{h}_l^T \mathbf{R}_{c_v,l}\mathbf{h}_l}, \ l = 1, 2, \ldots, L. \tag{3.11}$$

In general, $\text{oSNR}(\mathbf{h}_l) \neq \text{iSNR}_l$ except when $\mathbf{h}_l$ is a scalar. It is interesting to observe that the frequency-domain subband output SNR is not influenced by $H(j\omega)$.

We now define the fullband output SNRs for the frequency- and KLE-domain filters:

$$\text{oSNR}(H) = \frac{\int_{-\pi}^{\pi} |H(j\omega)|^2 \phi_x(\omega)d\omega}{\int_{-\pi}^{\pi} |H(j\omega)|^2 \phi_v(\omega)d\omega} \tag{3.12}$$

and

$$\text{oSNR}\left(\mathbf{h}_{1:L}\right) = \frac{\sum_{l=1}^{L} \mathbf{h}_l^T \mathbf{R}_{c_x,l}\mathbf{h}_l}{\sum_{l=1}^{L} \mathbf{h}_l^T \mathbf{R}_{c_v,l}\mathbf{h}_l}. \tag{3.13}$$

It is of great importance to find the complex gains $H(j\omega)$, $\omega \in (-\pi, \pi]$ and the FIR filters $\mathbf{h}_l$, $l = 1, 2, \ldots, L$, in such a way that $\text{oSNR}(H) \geq \text{iSNR}$ and $\text{oSNR}\left(\mathbf{h}_{1:L}\right) \geq \text{iSNR}$.

Property 3.1. We always have

$$\sum_{l=1}^{L} \text{iSNR}_l \geq \text{iSNR}, \tag{3.14}$$

$$\sum_{l=1}^{L} \text{oSNR}(\mathbf{h}_l) \geq \text{oSNR}\left(\mathbf{h}_{1:L}\right). \tag{3.15}$$

24   3 Performance Measures

This means that the aggregation of the subband SNRs is greater than or equal to the fullband SNR.

Proof. The two previous inequalities can be shown by using the following inequality:

$$\frac{\sum_l a_l}{\sum_l b_l} = \sum_l \left(\frac{a_l}{b_l} \cdot \frac{b_l}{\sum_i b_i} \right) \leq \sum_l \frac{a_l}{b_l}, \tag{3.16}$$

where a_l and b_l are positive reals.

3.2 Noise-Reduction Factor

Another important measure in noise reduction is the noise-reduction factor, which quantifies the amount of noise being attenuated by the filter. With the time-domain formulation, this factor is defined as [11], [24]

$$\xi_{nr}(\mathbf{H}) = \frac{\text{tr}(\mathbf{R}_v)}{\text{tr}\left(\mathbf{H}\mathbf{R}_v\mathbf{H}^T\right)}. \tag{3.17}$$

By analogy to the above time-domain definition, we define the subband noise-reduction factors in the frequency and KLE domains as

$$\xi_{nr}[H(j\omega)] = \frac{\phi_v(\omega)}{|H(j\omega)|^2 \phi_v(\omega)}$$

$$= \frac{1}{|H(j\omega)|^2}, \quad \omega \in (-\pi, \pi] \tag{3.18}$$

and

$$\xi_{nr}(\mathbf{h}_l) = \frac{\mathbf{q}_l^T \mathbf{R}_v \mathbf{q}_l}{\mathbf{h}_l^T \mathbf{R}_{c_v,l}\mathbf{h}_l}, \quad l = 1, 2, \ldots, L. \tag{3.19}$$

The larger the value of $\xi_{nr}[H(j\omega)]$ [or $\xi_{nr}(\mathbf{h}_l)$], the more the noise is reduced at frequency ω (or subband l). After the filtering operation, the residual noise level at frequency ω (or subband l) is expected to be lower than that of the original noise level, therefore this factor should have a lower bound of 1.

The fullband noise-reduction factors for the frequency- and KLE-domain filters are

$$\xi_{nr}(H) = \frac{\int_{-\pi}^{\pi} \phi_v(\omega)d\omega}{\int_{-\pi}^{\pi} |H(j\omega)|^2 \phi_v(\omega)d\omega} \tag{3.20}$$

$$= \frac{\int_{-\pi}^{\pi} \phi_v(\omega)d\omega}{\int_{-\pi}^{\pi} \xi_{nr}^{-1}[H(j\omega)] \phi_v(\omega)d\omega}$$

and

$$\xi_{\mathrm{nr}}\left(\mathbf{h}_{1:L}\right) = \frac{\sum_{l=1}^{L} \mathbf{q}_l^T \mathbf{R}_v \mathbf{q}_l}{\sum_{l=1}^{L} \mathbf{h}_l^T \mathbf{R}_{c_v,l} \mathbf{h}_l} \tag{3.21}$$

$$= \frac{\sum_{l=1}^{L} \mathbf{q}_l^T \mathbf{R}_v \mathbf{q}_l}{\sum_{l=1}^{L} \xi_{\mathrm{nr}}^{-1}\left(\mathbf{h}_l\right) \mathbf{q}_l^T \mathbf{R}_v \mathbf{q}_l}.$$

The fullband noise-reduction factor for the frequency-domain (or KLE-domain) approach is the ratio of the energy of the noise over the weighted energy of the noise with the weighting $\xi_{\mathrm{nr}}^{-1}\left[H(j\omega)\right]$ [or $\xi_{\mathrm{nr}}^{-1}\left(\mathbf{h}_l\right)$]. Same as in (3.18) [or (3.19)], $\xi_{\mathrm{nr}}(H)$ [or $\xi_{\mathrm{nr}}\left(\mathbf{h}_{1:L}\right)$] is expected to be lower bounded by 1. Indeed, if $\xi_{\mathrm{nr}}\left[H(j\omega)\right] \geq 1$, $\forall\omega$ [or $\xi_{\mathrm{nr}}\left(\mathbf{h}_l\right) \geq 1$, $\forall l$], we deduce from (3.20) [or (3.21)] that $\xi_{\mathrm{nr}}(H) \geq 1$ [or $\xi_{\mathrm{nr}}\left(\mathbf{h}_{1:L}\right) \geq 1$].

Property 3.2. We always have

$$\sum_{l=1}^{L} \xi_{\mathrm{nr}}\left(\mathbf{h}_l\right) \geq \xi_{\mathrm{nr}}\left(\mathbf{h}_{1:L}\right). \tag{3.22}$$

This means that the aggregation of the subband noise-reduction factors in the KLE domain is greater than or equal to the fullband noise-reduction factor.

Proof. This inequality can be shown by using (3.16).

3.3 Speech-Distortion Index

The filtering operation adds distortion to the speech signal. In order to evaluate the amount of speech distortion, the concept of speech-distortion index has been introduced in [11], [24]. With the time-domain model, the speech-distortion index is defined as

$$\begin{aligned} \upsilon_{\mathrm{sd}}\left(\mathbf{H}\right) &= \frac{E\left\{\left[\mathbf{x}_{\mathrm{F}}(k) - \mathbf{x}(k)\right]^T \left[\mathbf{x}_{\mathrm{F}}(k) - \mathbf{x}(k)\right]\right\}}{E\left[\mathbf{x}^T(k)\mathbf{x}(k)\right]} \\ &= \frac{E\left\{\left[\mathbf{H}\mathbf{x}(k) - \mathbf{x}(k)\right]^T \left[\mathbf{H}\mathbf{x}(k) - \mathbf{x}(k)\right]\right\}}{\mathrm{tr}\left(\mathbf{R}_x\right)} \\ &= \frac{\mathrm{tr}\left[(\mathbf{H} - \mathbf{I})\mathbf{R}_x(\mathbf{H} - \mathbf{I})^T\right]}{\mathrm{tr}\left(\mathbf{R}_x\right)}, \end{aligned} \tag{3.23}$$

where $\mathbf{I}$ is the identity matrix. We can extend this definition to other domains. Indeed, we define the subband speech-distortion indices in the frequency and KLE domains as

26   3 Performance Measures

$$v_{\mathrm{sd}}\left[H(j\omega)\right] = \frac{E\left[\left|H(j\omega)X(j\omega) - X(j\omega)\right|^2\right]}{\phi_x(\omega)}$$

$$= \left|1 - H(j\omega)\right|^2, \ \omega \in (-\pi, \pi] \tag{3.24}$$

and

$$v_{\mathrm{sd}}\left(\mathbf{h}_l\right) = \frac{E\left\{\left[\mathbf{h}_l^T \mathbf{c}_{x,l}(k) - c_{x,l}(k)\right]^2\right\}}{\lambda_l}, \ l = 1, 2, \ldots, L. \tag{3.25}$$

The subband speech-distortion index has a lower bound of 0 and an upper bound of 1 for optimal filters to be derived later in this book. The higher the value of $v_{\mathrm{sd}}\left[H(j\omega)\right]$ [or $v_{\mathrm{sd}}\left(\mathbf{h}_l\right)$], the more the speech is distorted at frequency ω (or subband l).

The fullband speech-distortion indices for the frequency- and KLE-domain filters are

$$v_{\mathrm{sd}}(H) = \frac{\int_{-\pi}^{\pi} E\left[\left|H(j\omega)X(j\omega) - X(j\omega)\right|^2\right] d\omega}{\int_{-\pi}^{\pi} \phi_x(\omega) d\omega}$$

$$= \frac{\int_{-\pi}^{\pi} \phi_x(\omega) \left|1 - H(j\omega)\right|^2 d\omega}{\int_{-\pi}^{\pi} \phi_x(\omega) d\omega}$$

$$= \frac{\int_{-\pi}^{\pi} v_{\mathrm{sd}}\left[H(j\omega)\right] \phi_x(\omega) d\omega}{\int_{-\pi}^{\pi} \phi_x(\omega) d\omega} \tag{3.26}$$

and

$$v_{\mathrm{sd}}\left(\mathbf{h}_{1:L}\right) = \frac{\sum_{l=1}^{L} E\left\{\left[\mathbf{h}_l^T \mathbf{c}_{x,l}(k) - c_{x,l}(k)\right]^2\right\}}{\sum_{l=1}^{L} \lambda_l}$$

$$= \frac{\sum_{l=1}^{L} v_{\mathrm{sd}}\left(\mathbf{h}_l\right) \lambda_l}{\sum_{l=1}^{L} \lambda_l}. \tag{3.27}$$

Equation (3.26) [or (3.27)] is the ratio of the weighted energy of the speech with the weighting $v_{\mathrm{sd}}\left[H(j\omega)\right]$ [or $v_{\mathrm{sd}}\left(\mathbf{h}_l\right)$] over the energy of the speech. If $v_{\mathrm{sd}}\left[H(j\omega)\right] \leq 1, \ \forall \omega$ [or $v_{\mathrm{sd}}\left(\mathbf{h}_l\right) \leq 1, \ \forall l$], we see from (3.26) [or (3.27)] that $v_{\mathrm{sd}}(H) \leq 1$ [or $v_{\mathrm{sd}}\left(\mathbf{h}_{1:L}\right) \leq 1$].

Property 3.3. We always have

$$\sum_{l=1}^{L} v_{\mathrm{sd}}\left(\mathbf{h}_l\right) \geq v_{\mathrm{sd}}\left(\mathbf{h}_{1:L}\right). \tag{3.28}$$

This means that the aggregation of the subband speech-distortion indices in the KLE domain is greater than or equal to the fullband speech-distortion index.

Proof. Easy to show by using the inequality (3.16).

3.4 Speech-Reduction Factor

This measure is somewhat similar to the noise-reduction factor. Since the noise is reduced by the filtering operation, so is the speech. This speech reduction implies, in general, speech distortion.

With the time-domain formulation, the speech-reduction factor is defined as

$$\xi_{\text{sr}}(\mathbf{H}) = \frac{\text{tr}(\mathbf{R}_x)}{\text{tr}\left(\mathbf{H}\mathbf{R}_x\mathbf{H}^T\right)}. \tag{3.29}$$

We define the subband speech-reduction factors in the frequency and KLE domains as

$$\xi_{\text{sr}}[H(j\omega)] = \frac{\phi_x(\omega)}{|H(j\omega)|^2\phi_x(\omega)}$$
$$= \frac{1}{|H(j\omega)|^2}, \ \omega \in (-\pi, \pi] \tag{3.30}$$

and

$$\xi_{\text{sr}}(\mathbf{h}_l) = \frac{\lambda_l}{\mathbf{h}_l^T\mathbf{R}_{c_x,l}\mathbf{h}_l}, \ l = 1, 2, \ldots, L. \tag{3.31}$$

The larger the value of $\xi_{\text{sr}}[H(j\omega)]$ [or $\xi_{\text{sr}}(\mathbf{h}_l)$], the more the speech is reduced at frequency ω (or subband l). After the filtering operation, the speech level at frequency ω (or subband l) is typically lower than that of the original speech level, therefore this factor should have a lower bound of 1.

The fullband speech-reduction factors for the frequency- and KLE-domain filters are

$$\xi_{\text{sr}}(H) = \frac{\int_{-\pi}^{\pi}\phi_x(\omega)d\omega}{\int_{-\pi}^{\pi}|H(j\omega)|^2\phi_x(\omega)d\omega} \tag{3.32}$$
$$= \frac{\int_{-\pi}^{\pi}\phi_x(\omega)d\omega}{\int_{-\pi}^{\pi}\xi_{\text{sr}}^{-1}[H(j\omega)]\phi_x(\omega)d\omega}$$

and

$$\xi_{\text{sr}}(\mathbf{h}_{1:L}) = \frac{\sum_{l=1}^{L}\lambda_l}{\sum_{l=1}^{L}\mathbf{h}_l^T\mathbf{R}_{c_x,l}\mathbf{h}_l} \tag{3.33}$$
$$= \frac{\sum_{l=1}^{L}\lambda_l}{\sum_{l=1}^{L}\xi_{\text{sr}}^{-1}(\mathbf{h}_l)\lambda_l}.$$

The fullband speech-reduction factor for the frequency domain (or KLE domain) approach is the ratio of the energy of the speech over the weighted

28   3 Performance Measures

energy of the speech with the weighting $\xi_{\mathrm{sr}}^{-1}[H(j\omega)]$ [or $\xi_{\mathrm{sr}}^{-1}(\mathbf{h}_l)$]. Same as in (3.30) [or (3.31)], $\xi_{\mathrm{sr}}(H)$ [or $\xi_{\mathrm{sr}}(\mathbf{h}_{1:L})$] is expected to be lower bounded by 1. Indeed, if $\xi_{\mathrm{sr}}[H(j\omega)] \geq 1$, $\forall\omega$ [or $\xi_{\mathrm{sr}}(\mathbf{h}_l) \geq 1$, $\forall l$], we deduce from (3.32) [or (3.33)] that $\xi_{\mathrm{sr}}(H) \geq 1$ [or $\xi_{\mathrm{sr}}(\mathbf{h}_{1:L}) \geq 1$].

Property 3.4. We always have

$$\sum_{l=1}^{L} \xi_{\mathrm{sr}}(\mathbf{h}_l) \geq \xi_{\mathrm{sr}}(\mathbf{h}_{1:L}).\tag{3.34}$$

This means that the aggregation of the subband speech-reduction factors in the KLE domain is greater than or equal to the fullband speech-reduction factor.

Proof. This inequality can be easily shown by using (3.16).

3.5 Discussion

We have presented several measures in this chapter. It is fair to ask whether all these measures are relevant and could give us a good idea on how a specific linear filter will behave in terms of reducing the noise and distorting the desired signal. The input SNR (independent of the filtering operation) is certainly a measure of great importance from both signal processing and perception points of view. Indeed, the human ear can do a pretty good job in judging the SNR values when we listen to some audio signals that are corrupted by additive noise with different SNRs. The subband input SNR, which is a narrowband definition, is less obvious to understand in our context since we usually do not listen to narrowband signals. But this measure is relevant since it is closely related to the input SNR. The output SNR (computed after the processing is done in the time domain) and the fullband output SNR (computed after the processing is done in the transform domain) are also very reliable. They tell us indeed whether the SNR is improved or not and by how much.

The noise-reduction factor is a relative measure because it does depend on the distortion (or reduction) of the desired signal. It is possible in some situations to have this factor much larger than 1 and yet the SNR has not been improved. So this measure has to be handled with care. The speech-distortion index and speech-reduction factor are very rough measures of distortion. Refined measures able to detect different kinds of distortions are more complicated to derive and much more research is needed in this topic. Nevertheless, four of the (fullband) measures are simply related as shown in the following property:

Property 3.5. We always have

$$\frac{\mathrm{oSNR}(\mathbf{H})}{\mathrm{iSNR}} = \frac{\xi_{\mathrm{nr}}(\mathbf{H})}{\xi_{\mathrm{sr}}(\mathbf{H})}.\tag{3.35}$$

Proof. This is easy to see by combining expressions (3.1), (3.9), (3.17), and (3.29). $\blacksquare$

From (3.35) we observe that oSNR(**H**) > iSNR if and only if $\xi_{nr}(\mathbf{H}) > \xi_{sr}(\mathbf{H})$. So is it possible that with a judicious choice of the filtering matrix, **H**, we can have $\xi_{nr}(\mathbf{H}) > \xi_{sr}(\mathbf{H})$? The answer is yes. A generally rough and intuitive justification to this answer is quite simple: improvement of the SNR is due to the fact that speech signals are partly predictable. In this situation, **H** is a kind of a complex predictor or interpolation matrix and as a result, $\xi_{sr}(\mathbf{H})$ can be close to 1 while $\xi_{nr}(\mathbf{H})$ can be much larger than 1. This observation is essential in order to understand how noise reduction happens when it happens.

We have similar relations to (3.35) for the filters in the frequency and KLE domains:

$$\frac{\text{oSNR}(H)}{\text{iSNR}} = \frac{\xi_{nr}(H)}{\xi_{sr}(H)}, \tag{3.36}$$

$$\frac{\text{oSNR}(\mathbf{h}_{1:L})}{\text{iSNR}} = \frac{\xi_{nr}(\mathbf{h}_{1:L})}{\xi_{sr}(\mathbf{h}_{1:L})}. \tag{3.37}$$

Even though the complex gains $H(j\omega)$, $\omega \in (-\pi, \pi]$ and the FIR filters $\mathbf{h}_l$, $l = 1, 2, \ldots, L$, may be interpreted differently than the time-domain filtering matrix **H** (since the formers act mostly like gain functions), the objective stays the same.

4

Mean-Squared Error Criterion

Error criteria play a critical role in deriving optimal noise reduction filters. Although many different criteria can be defined, the mean-squared error (MSE) is, by far, the most used one because of its simplicity in terms of deriving useful filters. This chapter presents relevant error signals from which meaningful MSE criteria are built in the different domains.

4.1 In the Time Domain

We define the error signal vector between the estimated and desired signals as

$$\mathbf{e}(k) = \mathbf{z}(k) - \mathbf{x}(k) \tag{4.1}$$
$$= \mathbf{H}\mathbf{y}(k) - \mathbf{x}(k),$$

which can also be written as the sum of two error signal vectors:

$$\mathbf{e}(k) = \mathbf{e}_x(k) + \mathbf{e}_v(k), \tag{4.2}$$

where

$$\mathbf{e}_x(k) = (\mathbf{H} - \mathbf{I})\,\mathbf{x}(k) \tag{4.3}$$

is the speech distortion due to the linear transformation, and

$$\mathbf{e}_v(k) = \mathbf{H}\mathbf{v}(k) \tag{4.4}$$

represents the residual noise [49].

Having defined the error signal, we can now write the MSE criterion:

$$J(\mathbf{H}) = \mathrm{tr}\left\{ E\left[\mathbf{e}(k)\mathbf{e}^T(k) \right] \right\} \tag{4.5}$$
$$= \mathrm{tr}\left(\mathbf{R}_x \right) + \mathrm{tr}\left(\mathbf{H}\mathbf{R}_y\mathbf{H}^T \right) - 2\mathrm{tr}\left(\mathbf{H}\mathbf{R}_{yx} \right)$$
$$= \mathrm{tr}\left(\mathbf{R}_x \right) + \mathrm{tr}\left(\mathbf{H}\mathbf{R}_y\mathbf{H}^T \right) - 2\mathrm{tr}\left(\mathbf{H}\mathbf{R}_x \right),$$

J. Benesty et al., *Noise Reduction in Speech Processing*, Springer Topics in Signal Processing 2,
DOI 10.1007/978-3-642-00296-0_4, © Springer-Verlag Berlin Heidelberg 2009

32   4 Mean-Squared Error Criterion

where

$$\mathbf{R}_{yx} = E\left[\mathbf{y}(k)\mathbf{x}^T(k)\right]$$

is the cross-correlation matrix between the observation and desired signals, which can also be expressed as

$$\mathbf{R}_{yx} = \mathbf{R}_x$$

since $\mathbf{R}_{vx} = E\left[\mathbf{v}(k)\mathbf{x}^T(k)\right] = \mathbf{0}$ [$x(k)$ and $v(k)$ are assumed to be uncorrelated]. Similarly, using the uncorrelation assumption, expression (4.5) can be structured in terms of two MSEs, i.e.,

$$\begin{aligned} J\left(\mathbf{H}\right) &= \text{tr}\left\{E\left[\mathbf{e}_x(k)\mathbf{e}_x^T(k)\right]\right\} + \text{tr}\left\{E\left[\mathbf{e}_v(k)\mathbf{e}_v^T(k)\right]\right\} \qquad (4.6)\\ &= J_x\left(\mathbf{H}\right) + J_v\left(\mathbf{H}\right). \end{aligned}$$

For the particular transformation $\mathbf{H} = \mathbf{I}$ (the identity matrix), we get

$$J\left(\mathbf{I}\right) = \text{tr}\left(\mathbf{R}_v\right), \qquad (4.7)$$

so there will be neither noise reduction nor speech distortion. Using this particular case of the MSE, we define the normalized MSE (NMSE) as

$$\begin{aligned} \tilde{J}\left(\mathbf{H}\right) &= \frac{J\left(\mathbf{H}\right)}{J\left(\mathbf{I}\right)}\\ &= \text{iSNR} \cdot v_{\text{sd}}\left(\mathbf{H}\right) + \frac{1}{\xi_{\text{nr}}\left(\mathbf{H}\right)}, \qquad (4.8) \end{aligned}$$

where

$$v_{\text{sd}}\left(\mathbf{H}\right) = \frac{J_x\left(\mathbf{H}\right)}{\text{tr}\left(\mathbf{R}_x\right)}, \qquad (4.9)$$

$$\xi_{\text{nr}}\left(\mathbf{H}\right) = \frac{\text{tr}\left(\mathbf{R}_v\right)}{J_v\left(\mathbf{H}\right)}. \qquad (4.10)$$

This shows the connection between the NMSE and the speech-distortion index and the noise-reduction factor defined in Chapter 3.

4.2 In the Frequency Domain

We define the error signal between the estimated and desired signals at frequency ω as

$$\begin{aligned} \mathcal{E}(j\omega) &= Z(j\omega) - X(j\omega) \qquad (4.11)\\ &= H(j\omega)Y(j\omega) - X(j\omega). \end{aligned}$$

This error can also be put into the form:

$$\mathcal{E}(j\omega) = \mathcal{E}_x(j\omega) + \mathcal{E}_v(j\omega), \tag{4.12}$$

where

$$\mathcal{E}_x(j\omega) = [H(j\omega) - 1]\,X(j\omega) \tag{4.13}$$

is the speech distortion due to the complex filter, and

$$\mathcal{E}_v(j\omega) = H(j\omega)V(j\omega) \tag{4.14}$$

represents the residual noise.

The frequency-domain (or subband) MSE is then

$$\begin{aligned}
J\left[H(j\omega)\right] &= E\left[|\mathcal{E}(j\omega)|^2\right] \tag{4.15}\\
&= \phi_x(\omega) + |H(j\omega)|^2\,\phi_y(\omega) - 2\mathcal{R}\left[H(j\omega)\phi_{yx}(j\omega)\right]\\
&= \phi_x(\omega) + |H(j\omega)|^2\,\phi_y(\omega) - 2\mathcal{R}\left[H(j\omega)\phi_x(\omega)\right],
\end{aligned}$$

where $\mathcal{R}(\cdot)$ is the real part of a complex number and

$$\begin{aligned}
\phi_{yx}(j\omega) &= E\left[Y(j\omega)X^*(j\omega)\right]\\
&= \phi_x(\omega)
\end{aligned}$$

is the cross-spectrum between the observation and speech signals. The subband MSE is also

$$\begin{aligned}
J\left[H(j\omega)\right] &= E\left[|\mathcal{E}_x(j\omega)|^2\right] + E\left[|\mathcal{E}_v(j\omega)|^2\right] \tag{4.16}\\
&= J_x\left[H(j\omega)\right] + J_v\left[H(j\omega)\right].
\end{aligned}$$

For the particular gain $H(j\omega) = 1$, $\forall\omega$, we get

$$J(1) = \phi_v(\omega), \tag{4.17}$$

so there will be neither noise reduction nor speech distortion. With this particular case of the MSE, we define the frequency-domain (or subband) NMSE as

$$\begin{aligned}
\tilde{J}\left[H(j\omega)\right] &= \frac{J\left[H(j\omega)\right]}{J(1)}\\
&= \mathrm{iSNR}(\omega)\cdot v_{\mathrm{sd}}\left[H(j\omega)\right] + \frac{1}{\xi_{\mathrm{nr}}\left[H(j\omega)\right]}, \tag{4.18}
\end{aligned}$$

where

$$v_{\mathrm{sd}}\left[H(j\omega)\right] = \frac{J_x\left[H(j\omega)\right]}{\phi_x(\omega)}, \tag{4.19}$$

$$\xi_{\mathrm{nr}}\left[H(j\omega)\right] = \frac{\phi_v(\omega)}{J_v\left[H(j\omega)\right]}. \tag{4.20}$$

34   4 Mean-Squared Error Criterion

The frequency-domain NMSE depends explicitly on the subband speech-distortion index and the subband noise-reduction factor.

Sometimes, it is also important to examine the MSE from the fullband point of view. We define the fullband MSE and fullband NMSE as

$$
\begin{aligned}
J(H) &= \frac{1}{2\pi} \int_{-\pi}^{\pi} J\left[H(j\omega)\right] d\omega \\
&= \frac{1}{2\pi} \int_{-\pi}^{\pi} \phi_x(\omega) \left|1 - H(j\omega)\right|^2 d\omega + \frac{1}{2\pi} \int_{-\pi}^{\pi} \phi_v(\omega) \left|H(j\omega)\right|^2 d\omega \\
&= J_x(H) + J_v(H)
\end{aligned}
\tag{4.21}
$$

and

$$
\begin{aligned}
\tilde{J}(H) &= 2\pi \frac{J(H)}{\int_{-\pi}^{\pi} \phi_v(\omega) d\omega} \\
&= \frac{\int_{-\pi}^{\pi} \phi_x(\omega) \left|1 - H(j\omega)\right|^2 d\omega}{\int_{-\pi}^{\pi} \phi_v(\omega) d\omega} + \frac{\int_{-\pi}^{\pi} \phi_v(\omega) \left|H(j\omega)\right|^2 d\omega}{\int_{-\pi}^{\pi} \phi_v(\omega) d\omega} \\
&= \text{iSNR} \cdot \upsilon_{\text{sd}}(H) + \frac{1}{\xi_{\text{nr}}(H)},
\end{aligned}
\tag{4.22}
$$

where

$$
\upsilon_{\text{sd}}(H) = \frac{J_x(H)}{\int_{-\pi}^{\pi} \phi_x(\omega) d\omega},
\tag{4.23}
$$

$$
\xi_{\text{nr}}(H) = \frac{\int_{-\pi}^{\pi} \phi_v(\omega) d\omega}{J_v(H)}.
\tag{4.24}
$$

Again, we see clearly that the fullband NMSE depends explicitly on the fullband speech-distortion index and the fullband noise-reduction factor.

4.3 In the KLE Domain

In the KLE domain, the error signal between the estimated and desired signals in the subband l is

$$
\begin{aligned}
e_l(k) &= c_{z,l}(k) - c_{x,l}(k) \\
&= \mathbf{h}_l^T \mathbf{c}_{y,l}(k) - c_{x,l}(k),
\end{aligned}
\tag{4.25}
$$

which can also be written as the sum of two error signals:

$$
e_l(k) = e_{x,l}(k) + e_{v,l}(k),
\tag{4.26}
$$

where

$$e_{x,l}(k) = \mathbf{h}_l^T \mathbf{c}_{x,l}(k) - c_{x,l}(k) \tag{4.27}$$

is the speech distortion due to the FIR filter, and

$$e_{v,l}(k) = \mathbf{h}_l^T \mathbf{c}_{v,l}(k) \tag{4.28}$$

represents the residual noise.

From the error signal (4.25), we give the corresponding KLE-domain (or subband) MSE criterion:

$$\begin{aligned} J(\mathbf{h}_l) &= E\left[e_l^2(k)\right] \\ &= \lambda_l + \mathbf{h}_l^T \mathbf{R}_{c_y,l} \mathbf{h}_l - 2\mathbf{h}_l^T \mathbf{r}_{c_y c_x,l}, \end{aligned} \tag{4.29}$$

where

$$\begin{aligned} \mathbf{R}_{c_y,l} &= E\left[\mathbf{c}_{y,l}(k)\mathbf{c}_{y,l}^T(k)\right], \\ \mathbf{r}_{c_y c_x,l} &= E\left[\mathbf{c}_{y,l}(k)c_{x,l}(k)\right], \end{aligned}$$

are the correlation matrix of the signal $c_{y,l}(k)$ and cross-correlation vector between the signals $c_{y,l}(k)$ and $c_{x,l}(k)$, respectively. Expression (4.29) can be structured in a different way:

$$\begin{aligned} J(\mathbf{h}_l) &= E\left[e_{x,l}^2(k)\right] + E\left[e_{v,l}^2(k)\right] \\ &= J_x(\mathbf{h}_l) + J_v(\mathbf{h}_l). \end{aligned} \tag{4.30}$$

For the particular FIR filter

$$\mathbf{i}_l = \begin{bmatrix} 1 & 0 & \cdots & 0 \end{bmatrix}^T$$

of length L_l, we get

$$\begin{aligned} J(\mathbf{i}_l) &= E\left[c_{v,l}^2(k)\right] \\ &= \mathbf{q}_l^T \mathbf{R}_v \mathbf{q}_l, \end{aligned} \tag{4.31}$$

so there will be neither noise reduction nor speech distortion. Using this particular case of the MSE, we define the KLE-domain (or subband) NMSE as

$$\begin{aligned} \tilde{J}(\mathbf{h}_l) &= \frac{J(\mathbf{h}_l)}{J(\mathbf{i}_l)} \\ &= \mathrm{iSNR}_l \cdot \upsilon_{\mathrm{sd}}(\mathbf{h}_l) + \frac{1}{\xi_{\mathrm{nr}}(\mathbf{h}_l)}, \end{aligned} \tag{4.32}$$

where

$$\upsilon_{\mathrm{sd}}(\mathbf{h}_l) = \frac{J_x(\mathbf{h}_l)}{\lambda_l}, \tag{4.33}$$

$$\xi_{\mathrm{nr}}(\mathbf{h}_l) = \frac{\mathbf{q}_l^T \mathbf{R}_v \mathbf{q}_l}{J_v(\mathbf{h}_l)}. \tag{4.34}$$

The KLE-domain NMSE depends explicitly on the subband speech-distortion index and the subband noise-reduction factor.

We define the fullband MSE and fullband NMSE as

$$J\left(\mathbf{h}_{1:L}\right) = \frac{1}{L}\sum_{l=1}^{L} J\left(\mathbf{h}_l\right) \tag{4.35}$$

$$= \frac{1}{L}\sum_{l=1}^{L}(\mathbf{h}_l - \mathbf{i}_l)^T \mathbf{R}_{c_x,l}(\mathbf{h}_l - \mathbf{i}_l) + \frac{1}{L}\sum_{l=1}^{L}\mathbf{h}_l^T \mathbf{R}_{c_v,l}\mathbf{h}_l$$

$$= J_x\left(\mathbf{h}_{1:L}\right) + J_v\left(\mathbf{h}_{1:L}\right)$$

and

$$\tilde{J}\left(\mathbf{h}_{1:L}\right) = L\frac{J\left(\mathbf{h}_{1:L}\right)}{\sum_{l=1}^{L}\mathbf{q}_l^T \mathbf{R}_v\mathbf{q}_l} \tag{4.36}$$

$$= \frac{\sum_{l=1}^{L}(\mathbf{h}_l - \mathbf{i}_l)^T \mathbf{R}_{c_x,l}(\mathbf{h}_l - \mathbf{i}_l)}{\sum_{l=1}^{L}\mathbf{q}_l^T \mathbf{R}_v\mathbf{q}_l} + \frac{\sum_{l=1}^{L}\mathbf{h}_l^T \mathbf{R}_{c_v,l}\mathbf{h}_l}{\sum_{l=1}^{L}\mathbf{q}_l^T \mathbf{R}_v\mathbf{q}_l}$$

$$= \text{iSNR} \cdot v_{\text{sd}}(\mathbf{h}_{1:L}) + \frac{1}{\xi_{\text{nr}}(\mathbf{h}_{1:L})},$$

where

$$v_{\text{sd}}(\mathbf{h}_{1:L}) = \frac{J_x\left(\mathbf{h}_{1:L}\right)}{\sum_{l=1}^{L}\lambda_l}, \tag{4.37}$$

$$\xi_{\text{nr}}(\mathbf{h}_{1:L}) = \frac{\sum_{l=1}^{L}\mathbf{q}_l^T \mathbf{R}_v\mathbf{q}_l}{J_v\left(\mathbf{h}_{1:L}\right)}. \tag{4.38}$$

Same as for the frequency-domain approach, the fullband NMSE with the KLE depends explicitly on the fullband speech-distortion index and the fullband noise-reduction factor.

4.4 Summary

In this chapter, we have presented the MSE and NMSE criteria in the time, frequency, and KLE domains. It has been demonstrated that the (subband or fullband) NMSE in all domains depends explicitly on the input SNR, speech-distortion index, and noise-reduction factor, which makes the NMSE not only useful in deriving different optimal filters, but also powerful in analyzing noise reduction performance. In noise reduction, it is also very important to directly compare the output SNR with the input SNR. This is essential in order to tell whether the filter is really able to achieve noise reduction. However, the MSE and NMSE do not show explicitly the output SNR even though intuitively (and implicitly) they should depend on it. In the next chapter, we will present another criterion, called the Pearson correlation coefficient, in which the output SNR appears naturally.

5

Pearson Correlation Coefficient

This chapter develops several forms of the Pearson correlation coefficient in the different domains. This coefficient can be used as an optimization criterion to derive different optimal noise reduction filters [14], but is even more useful for analyzing these optimal filters for their noise reduction performance.

5.1 Correlation Coefficient Between Two Random Variables

Let a and b be two zero-mean real-valued random variables. The Pearson correlation coefficient (PCC) is defined as[1] [41], [98], [105]

$$\rho(a, b) = \frac{E(ab)}{\sigma_a \sigma_b}, \tag{5.1}$$

where $E(ab)$ is the cross-correlation between a and b, and $\sigma_a^2 = E(a^2)$ and $\sigma_b^2 = E(b^2)$ are the variances of the signals a and b, respectively. In the rest, it will be more convenient to work with the squared Pearson correlation coefficient (SPCC):

$$\rho^2(a, b) = \frac{E^2(ab)}{\sigma_a^2 \sigma_b^2}. \tag{5.2}$$

One of the most important properties of the SPCC is that

$$0 \le \rho^2(a, b) \le 1. \tag{5.3}$$

The SPCC gives an indication on the strength of the linear relationship between the two random variables a and b. If $\rho^2(a, b) = 0$, then a and b are said

---

[1] This correlation coefficient is named after Karl Pearson who described many of its properties. Actually to be precise, the true definition of the PCC is when the expectation operator in (5.1) is replaced by a sum over the samples.

J. Benesty et al., *Noise Reduction in Speech Processing*, Springer Topics in Signal Processing 2, DOI 10.1007/978-3-642-00296-0_5, © Springer-Verlag Berlin Heidelberg 2009

38   5 Pearson Correlation Coefficient

to be uncorrelated. The closer the value of $\rho^2(a, b)$ is to 1, the stronger the correlation between the two variables. If the two variables are independent, then $\rho^2(a, b) = 0$. But the converse is not true because the SPCC detects only *linear* dependencies between the two variables a and b. For a non-linear dependency, the SPCC may be equal to zero. However, in the special case when a and b are jointly normal, "independent" is equivalent to "uncorrelated."

5.2 Correlation Coefficient Between Two Random Vectors

We can generalize the idea of the SPCC between two random variables to two random vectors. Indeed, let

$$\mathbf{a} = \begin{bmatrix} a_1 & a_2 & \cdots & a_L \end{bmatrix}^T,$$
$$\mathbf{b} = \begin{bmatrix} b_1 & b_2 & \cdots & b_L \end{bmatrix}^T,$$

be two zero-mean real-valued random vectors of length L. We define the SPCC between $\mathbf{a}$ and $\mathbf{b}$ as

$$\rho^2(\mathbf{a}, \mathbf{b}) = \frac{E^2(\mathbf{a}^T\mathbf{b})}{E(\mathbf{a}^T\mathbf{a}) E(\mathbf{b}^T\mathbf{b})}. \tag{5.4}$$

Let $\mathbf{\Pi}_a$ and $\mathbf{\Pi}_b$ be two permutation matrices. If $\mathbf{\Pi}_a = \mathbf{\Pi}_b$ then $\rho^2(\mathbf{\Pi}_a\mathbf{a}, \mathbf{\Pi}_a\mathbf{b}) = \rho^2(\mathbf{a}, \mathbf{b})$. In general, if $\mathbf{\Pi}_a \neq \mathbf{\Pi}_b$, we have $\rho^2(\mathbf{\Pi}_a\mathbf{a}, \mathbf{\Pi}_b\mathbf{b}) \neq \rho^2(\mathbf{a}, \mathbf{b})$.

Property 5.1. We always have

$$0 \leq \rho^2(\mathbf{a}, \mathbf{b}) \leq 1. \tag{5.5}$$

Proof. From the definition (5.4), it is clear that $\rho^2(\mathbf{a}, \mathbf{b}) \geq 0$. To show that $\rho^2(\mathbf{a}, \mathbf{b}) \leq 1$, let us define the positive quantity:

$$E\left[(\mathbf{a} - c\mathbf{b})^T(\mathbf{a} - c\mathbf{b})\right] \geq 0, \tag{5.6}$$

where c is a real number. The development of the previous expression gives:

$$E\left[(\mathbf{a} - c\mathbf{b})^T(\mathbf{a} - c\mathbf{b})\right] = E(\mathbf{a}^T\mathbf{a}) - 2cE(\mathbf{a}^T\mathbf{b}) + c^2 E(\mathbf{b}^T\mathbf{b}). \tag{5.7}$$

In particular, for

$$c = \frac{E(\mathbf{a}^T\mathbf{a})}{E(\mathbf{a}^T\mathbf{b})} \tag{5.8}$$

we get

$$E\left(\mathbf{a}^T\mathbf{a}\right) - 2E\left(\mathbf{a}^T\mathbf{a}\right) + \frac{E^2\left(\mathbf{a}^T\mathbf{a}\right)E\left(\mathbf{b}^T\mathbf{b}\right)}{E^2\left(\mathbf{a}^T\mathbf{b}\right)} \geq 0, \tag{5.9}$$

which implies that

$$\frac{E\left(\mathbf{a}^T\mathbf{a}\right)E\left(\mathbf{b}^T\mathbf{b}\right)}{E^2\left(\mathbf{a}^T\mathbf{b}\right)} \geq 1. \tag{5.10}$$

Therefore $\rho^2\left(\mathbf{a},\mathbf{b}\right) \leq 1$.

5.3 Frequency-Domain Versions

Let $A(j\omega)$ and $B(j\omega)$ be the DTFTs of the two zero-mean real-valued random variables a and b. We define the subband SPCC, which is also known as the magnitude squared coherence function (MSCF), between $A(j\omega)$ and $B(j\omega)$ at frequency ω as

$$\begin{aligned}
\left|\rho\left[A(j\omega), B(j\omega)\right]\right|^2 &= \frac{\left|E\left[A(j\omega)B^*(j\omega)\right]\right|^2}{E\left[\left|A(j\omega)\right|^2\right]E\left[\left|B(j\omega)\right|^2\right]} \\
&= \frac{\left|\phi_{ab}(j\omega)\right|^2}{\phi_a(\omega)\phi_b(\omega)},
\end{aligned} \tag{5.11}$$

where superscript $*$ denotes complex conjugation. It is clear that the subband SPCC always takes its values between 0 and 1.

We can generalize this idea to infinite vectors (containing all frequencies). We will refer to this definition as the fullband SPCC, which also takes its values between 0 and 1. In this case, we have

$$\begin{aligned}
\left|\rho\left(A, B\right)\right|^2 &= \frac{\left|E\left[\int_{-\pi}^{\pi} A(j\omega)B^*(j\omega)d\omega\right]\right|^2}{E\left[\int_{-\pi}^{\pi}\left|A(j\omega)\right|^2 d\omega\right]E\left[\int_{-\pi}^{\pi}\left|B(j\omega)\right|^2 d\omega\right]} \\
&= \frac{\left|\int_{-\pi}^{\pi}\phi_{ab}(j\omega)d\omega\right|^2}{\left[\int_{-\pi}^{\pi}\phi_a(\omega)d\omega\right]\left[\int_{-\pi}^{\pi}\phi_b(\omega)d\omega\right]} \\
&= \frac{E^2\left(ab\right)}{\sigma_a^2\sigma_b^2} \\
&= \rho^2\left(a, b\right).
\end{aligned} \tag{5.12}$$

5.4 KLE-Domain Versions

Let a and b be two zero-mean real-valued random variables and $c_{a,l}$ and $c_{b,l}$ their respective representations in the KLE domain and in the subband l. We

40   5 Pearson Correlation Coefficient

define the subband SPCC (or MSCF) between $c_{a,l}$ and $c_{b,l}$ in the subband l as

$$\rho^2\left(c_{a,l}, c_{b,l}\right) = \frac{E^2\left(c_{a,l} c_{b,l}\right)}{E\left(c_{a,l}^2\right) E\left(c_{b,l}^2\right)}. \tag{5.13}$$

The vector form or fullband SPCC is

$$\rho^2\left(\mathbf{c}_a, \mathbf{c}_b\right) = \frac{E^2\left(\mathbf{c}_a^T \mathbf{c}_b\right)}{E\left(\mathbf{c}_a^T \mathbf{c}_a\right) E\left(\mathbf{c}_b^T \mathbf{c}_b\right)}, \tag{5.14}$$

where $\mathbf{c}_a$ and $\mathbf{c}_b$ are two vectors of length L containing all the elements $c_{a,l}$ and $c_{b,l}$, $l = 1, 2, \ldots, L$, respectively. It is clear that

$$0 \leq \rho^2\left(c_a, c_b\right) \leq 1, \tag{5.15}$$

$$0 \leq \rho^2\left(\mathbf{c}_a, \mathbf{c}_b\right) \leq 1. \tag{5.16}$$

5.5 Summary

This chapter developed different forms of the so-called Pearson correlation coefficient in the time, frequency, and KLE domains. Each of these forms has many interesting properties, which are very useful not only for deriving, but also for analyzing optimal filters in the context of noise reduction. We will elaborate on these properties in the next chapter.

6

Fundamental Properties

Having introduced in the previous chapter several forms of the SPCC in different domains, we now study their fundamental properties. Armed with these properties it is possible in many cases to derive and analyze optimal noise reduction filters by a simple inspection of a certain form of the SPCC.

6.1 In the Time Domain

The SPCC between the desired and observation vectors $\mathbf{x}(k)$ and $\mathbf{y}(k)$ [as defined in (2.2)] is

$$
\begin{aligned}
\rho^2\left(\mathbf{x}, \mathbf{y}\right) &= \frac{E^2\left(\mathbf{x}^T \mathbf{y}\right)}{E\left(\mathbf{x}^T \mathbf{x}\right) E\left(\mathbf{y}^T \mathbf{y}\right)} \\
&= \frac{\mathrm{tr}(\mathbf{R}_x)}{\mathrm{tr}(\mathbf{R}_y)} \\
&= \frac{\mathrm{iSNR}}{1+\mathrm{iSNR}}.
\end{aligned}
\tag{6.1}
$$

So this SPCC says how much the observation signal is noisy. Indeed, a value of $\rho^2\left(\mathbf{x}, \mathbf{y}\right)$ close to 1 implies that the speech is largely dominant while a value of $\rho^2\left(\mathbf{x}, \mathbf{y}\right)$ close to 0 implies that the noise is largely dominant.

The SPCC between the two vectors $\mathbf{x}(k)$ and $\mathbf{z}(k)$ [as defined in (2.6)] is

$$
\begin{aligned}
\rho^2\left(\mathbf{x}, \mathbf{z}\right) &= \frac{\mathrm{tr}^2(\mathbf{H}\mathbf{R}_x)}{\mathrm{tr}(\mathbf{R}_x)\mathrm{tr}\left(\mathbf{H}\mathbf{R}_y\mathbf{H}^T\right)} \\
&= \frac{\mathrm{tr}^2(\mathbf{H}\mathbf{R}_x)}{\mathrm{tr}(\mathbf{R}_x)\mathrm{tr}\left(\mathbf{H}\mathbf{R}_x\mathbf{H}^T\right)} \cdot \frac{\mathrm{oSNR}(\mathbf{H})}{1+\mathrm{oSNR}(\mathbf{H})}.
\end{aligned}
\tag{6.2}
$$

Property 6.1. We have

J. Benesty et al., *Noise Reduction in Speech Processing*, Springer Topics in Signal Processing 2,
DOI 10.1007/978-3-642-00296-0_6, © Springer-Verlag Berlin Heidelberg 2009

42   6 Fundamental Properties

$$\rho^2\left(\mathbf{x}, \mathbf{z}\right) = \rho^2\left(\mathbf{x}, \mathbf{x}_{\mathrm{F}}\right) \cdot \rho^2\left(\mathbf{x}_{\mathrm{F}}, \mathbf{z}\right), \tag{6.3}$$

where

$$\rho^2\left(\mathbf{x}, \mathbf{x}_{\mathrm{F}}\right) = \frac{\mathrm{tr}^2(\mathbf{H}\mathbf{R}_x)}{\mathrm{tr}(\mathbf{R}_x)\mathrm{tr}\left(\mathbf{H}\mathbf{R}_x\mathbf{H}^T\right)} \tag{6.4}$$

and

$$\rho^2\left(\mathbf{x}_{\mathrm{F}}, \mathbf{z}\right) = \frac{\mathrm{oSNR}(\mathbf{H})}{1 + \mathrm{oSNR}(\mathbf{H})}. \tag{6.5}$$

The SPCC $\rho^2\left(\mathbf{x}, \mathbf{x}_{\mathrm{F}}\right)$ can be viewed as another definition of the speech-distortion index. If $\mathbf{H} = \mathbf{I}$ (no speech distortion) then $\rho^2\left(\mathbf{x}, \mathbf{x}_{\mathrm{F}}\right) = 1$. The closer the value of $\rho^2\left(\mathbf{x}, \mathbf{x}_{\mathrm{F}}\right)$ is to 0, the more the speech signal is distorted. The SPCC $\rho^2\left(\mathbf{x}_{\mathrm{F}}, \mathbf{z}\right)$ shows the SNR improvement and reaches its maximum when oSNR($\mathbf{H}$) is maximized. Intuitively, we would like to find a matrix $\mathbf{H}$ in such a way that the signal vector $\mathbf{z}(k)$ is as close as possible to the vector $\mathbf{x}(k)$; mathematically, this can be translated by maximizing the SPCC $\rho^2\left(\mathbf{x}, \mathbf{z}\right)$ or by minimizing $\rho^{-2}\left(\mathbf{x}, \mathbf{z}\right)$. Clearly, from this property [eq. (6.3)], the SNR improvement is made at the expense of speech distortion. Maximizing the output SNR is equivalent to maximizing the SPCC $\rho^2\left(\mathbf{x}_{\mathrm{F}}, \mathbf{z}\right)$. The filter derived from this optimization will introduce a large distortion to the desired signal since we try to make the signal vector $\mathbf{H}\mathbf{y}(k) = \mathbf{z}(k)$ as close as possible to the signal vector $\mathbf{H}\mathbf{x}(k) = \mathbf{x}_{\mathrm{F}}(k)$ [instead of $\mathbf{x}(k)$], which is a filtered version of $\mathbf{x}(k)$.

Now, if we take the natural logarithm of $\rho^{-2}\left(\mathbf{x}, \mathbf{z}\right)$, we find that

$$\ln\left[\rho^{-2}\left(\mathbf{x}, \mathbf{z}\right)\right] = \ln\left[\mathrm{tr}\left(\mathbf{R}_x\right)\right] + \ln\left[\mathrm{tr}\left(\mathbf{H}\mathbf{R}_y\mathbf{H}^T\right)\right] - \ln\left[\mathrm{tr}^2\left(\mathbf{H}\mathbf{R}_x\right)\right]. \tag{6.6}$$

It is instructive to compare this criterion to the MSE $J\left(\mathbf{H}\right)$ [eq. (4.5)].

Property 6.2. We have

$$\rho^2\left(\mathbf{x}, \mathbf{z}\right) \le \frac{\mathrm{oSNR}(\mathbf{H})}{1 + \mathrm{oSNR}(\mathbf{H})}, \tag{6.7}$$

with equality when $\mathbf{H} = \mathbf{I}$.

Proof. This property follows immediately from (6.3) since $\rho^2\left(\mathbf{x}, \mathbf{x}_{\mathrm{F}}\right) \le 1$.

Property 6.3. We have

$$\rho^2\left(\mathbf{x}_{\mathrm{F}}, \mathbf{y}\right) = \rho^2\left(\mathbf{x}, \mathbf{x}_{\mathrm{F}}\right) \cdot \rho^2\left(\mathbf{x}, \mathbf{y}\right). \tag{6.8}$$

Proof. Indeed

$$\rho^2\left(\mathbf{x}_{\mathrm{F}}, \mathbf{y}\right) = \frac{\mathrm{tr}^2(\mathbf{H}\mathbf{R}_x)}{\mathrm{tr}\left(\mathbf{H}\mathbf{R}_x\mathbf{H}^T\right)\mathrm{tr}(\mathbf{R}_y)}$$

$$= \frac{\mathrm{tr}^2(\mathbf{H}\mathbf{R}_x)}{\mathrm{tr}\left(\mathbf{H}\mathbf{R}_x\mathbf{H}^T\right)\mathrm{tr}(\mathbf{R}_x)} \cdot \frac{\mathrm{iSNR}}{1 + \mathrm{iSNR}}$$

$$= \rho^2\left(\mathbf{x}, \mathbf{x}_{\mathrm{F}}\right) \cdot \rho^2\left(\mathbf{x}, \mathbf{y}\right).$$

Property 6.4. We have

$$\rho^2\left(\mathbf{x}_{\mathrm{F}}, \mathbf{y}\right) \leq \frac{\mathrm{iSNR}}{1 + \mathrm{iSNR}}, \tag{6.9}$$

with equality when $\mathbf{H} = \mathbf{I}$.

Proof. This property follows immediately from (6.8) since $\rho^2\left(\mathbf{x}, \mathbf{x}_{\mathrm{F}}\right) \leq 1$.

The SPCC between the two vectors $\mathbf{v}(k)$ and $\mathbf{y}(k)$ is another way to see how much the microphone signal is affected by the noise. This SPCC is

$$\rho^2\left(\mathbf{v}, \mathbf{y}\right) = \frac{\mathrm{tr}(\mathbf{R}_v)}{\mathrm{tr}(\mathbf{R}_y)}$$

$$= \frac{1}{1 + \mathrm{iSNR}}. \tag{6.10}$$

So a value of $\rho^2\left(\mathbf{v}, \mathbf{y}\right)$ close to 1 implies that the noise is largely dominant while a value of $\rho^2\left(\mathbf{v}, \mathbf{y}\right)$ close to 0 implies that the speech is largely dominant.

Property 6.5. We have

$$\rho^2\left(\mathbf{x}, \mathbf{y}\right) + \rho^2\left(\mathbf{v}, \mathbf{y}\right) = 1 \tag{6.11}$$

and

$$\mathrm{iSNR} = \frac{\rho^2\left(\mathbf{x}, \mathbf{y}\right)}{\rho^2\left(\mathbf{v}, \mathbf{y}\right)}. \tag{6.12}$$

Proof. Easy to see from (6.1) and (6.10).

The input SNR is the ratio between two SPCCs and the sum of these two SPCCs is equal to 1. We can give another simple interpretation of Property 6.5. Let us define the complex number

$$\varrho\left(\mathbf{x}, \mathbf{v}\right) = \rho\left(\mathbf{x}, \mathbf{y}\right) + j\rho\left(\mathbf{v}, \mathbf{y}\right)$$

$$= \cos\theta + j\sin\theta, \tag{6.13}$$

where θ is the angle of $\varrho\left(\mathbf{x}, \mathbf{v}\right)$ for which the modulus is equal to 1. On the complex plane, this complex number is on the unit circle. Since $0 \leq \rho\left(\mathbf{x}, \mathbf{y}\right) \leq 1$ and $0 \leq \rho\left(\mathbf{v}, \mathbf{y}\right) \leq 1$, therefore $0 \leq \theta \leq \frac{\pi}{2}$. Hence

44   6 Fundamental Properties

$$\lim_{\theta \to 0} \text{iSNR} = \infty, \tag{6.14}$$

$$\lim_{\theta \to \frac{\pi}{2}} \text{iSNR} = 0. \tag{6.15}$$

In other words, $\varrho(\mathbf{x}, \mathbf{v})$ is purely real (resp. imaginary) if and only if the input SNR is equal to infinity (resp. zero). In this interpretation, we represent the input SNR by the angle θ for which it takes its values between 0 and $\pi/2$.

The SPCC between the two vectors $\mathbf{v}(k)$ and $\mathbf{z}(k)$ is

$$\begin{aligned}
\rho^2(\mathbf{v}, \mathbf{z}) &= \frac{\text{tr}^2(\mathbf{H}\mathbf{R}_v)}{\text{tr}(\mathbf{R}_v)\text{tr}\left(\mathbf{H}\mathbf{R}_y\mathbf{H}^T\right)} \\
&= \frac{\text{tr}^2(\mathbf{H}\mathbf{R}_v)}{\text{tr}(\mathbf{R}_v)\text{tr}\left(\mathbf{H}\mathbf{R}_v\mathbf{H}^T\right)} \cdot \frac{1}{1 + \text{oSNR}(\mathbf{H})}.
\end{aligned} \tag{6.16}$$

Property 6.6. We have

$$\rho^2(\mathbf{v}, \mathbf{z}) = \rho^2(\mathbf{v}, \mathbf{v}_{\text{F}}) \cdot \rho^2(\mathbf{v}_{\text{F}}, \mathbf{z}), \tag{6.17}$$

where

$$\rho^2(\mathbf{v}, \mathbf{v}_{\text{F}}) = \frac{\text{tr}^2(\mathbf{H}\mathbf{R}_v)}{\text{tr}(\mathbf{R}_v)\text{tr}\left(\mathbf{H}\mathbf{R}_v\mathbf{H}^T\right)} \tag{6.18}$$

and

$$\rho^2(\mathbf{v}_{\text{F}}, \mathbf{z}) = \frac{1}{1 + \text{oSNR}(\mathbf{H})}. \tag{6.19}$$

Property 6.7. We have

$$\rho^2(\mathbf{v}, \mathbf{z}) \le \frac{1}{1 + \text{oSNR}(\mathbf{H})}, \tag{6.20}$$

with equality when $\mathbf{H} = \mathbf{I}$.

Proof. This property follows immediately from (6.17) since $\rho^2(\mathbf{v}, \mathbf{v}_{\text{F}}) \le 1$.

Property 6.8. We have

$$\rho^2(\mathbf{v}_{\text{F}}, \mathbf{y}) = \rho^2(\mathbf{v}, \mathbf{v}_{\text{F}}) \cdot \rho^2(\mathbf{v}, \mathbf{y}). \tag{6.21}$$

Proof. Indeed

$$\begin{aligned}
\rho^2(\mathbf{v}_{\text{F}}, \mathbf{y}) &= \frac{\text{tr}^2(\mathbf{H}\mathbf{R}_v)}{\text{tr}\left(\mathbf{H}\mathbf{R}_v\mathbf{H}^T\right)\text{tr}(\mathbf{R}_y)} \\
&= \frac{\text{tr}^2(\mathbf{H}\mathbf{R}_v)}{\text{tr}\left(\mathbf{H}\mathbf{R}_v\mathbf{H}^T\right)\text{tr}(\mathbf{R}_v)} \cdot \frac{1}{1 + \text{iSNR}} \\
&= \rho^2(\mathbf{v}, \mathbf{v}_{\text{F}}) \cdot \rho^2(\mathbf{v}, \mathbf{y}).
\end{aligned}$$

6.1 In the Time Domain   45

Property 6.9. We have

$$\rho^2\left(\mathbf{v}_\mathrm{F}, \mathbf{y}\right) \le \frac{1}{1 + \mathrm{iSNR}}, \tag{6.22}$$

with equality when $\mathbf{H} = \mathbf{I}$.

Proof. This property follows immediately from (6.21) since $\rho^2\left(\mathbf{v}, \mathbf{v}_\mathrm{F}\right) \le 1$.

Property 6.10. We have

$$\rho^2\left(\mathbf{x}_\mathrm{F}, \mathbf{z}\right) + \rho^2\left(\mathbf{v}_\mathrm{F}, \mathbf{z}\right) = 1 \tag{6.23}$$

and

$$\mathrm{oSNR}(\mathbf{H}) = \frac{\rho^2\left(\mathbf{x}_\mathrm{F}, \mathbf{z}\right)}{\rho^2\left(\mathbf{v}_\mathrm{F}, \mathbf{z}\right)}. \tag{6.24}$$

Proof. Easy to see from (6.5) and (6.19).

The output SNR is the ratio between two SPCCs depending on $\mathbf{H}$ and the sum of these two SPCCs is equal to 1. Now, let us define the complex number after the filtering processing:

$$\begin{aligned}\varrho\left(\mathbf{x}_\mathrm{F}, \mathbf{v}_\mathrm{F}\right) &= \rho\left(\mathbf{x}_\mathrm{F}, \mathbf{z}\right) + j\rho\left(\mathbf{v}_\mathrm{F}, \mathbf{z}\right)\\ &= \cos\theta_\mathrm{F} + j\sin\theta_\mathrm{F}, \end{aligned} \tag{6.25}$$

where θ_F is the angle of $\varrho\left(\mathbf{x}_\mathrm{F}, \mathbf{v}_\mathrm{F}\right)$ for which the modulus is equal to 1. On the complex plane, this complex number is on the unit circle. Since $0 \le \rho\left(\mathbf{x}_\mathrm{F}, \mathbf{z}\right) \le 1$ and $0 \le \rho\left(\mathbf{v}_\mathrm{F}, \mathbf{z}\right) \le 1$, therefore $0 \le \theta_\mathrm{F} \le \frac{\pi}{2}$. Hence

$$\lim_{\theta_\mathrm{F} \to 0} \mathrm{oSNR}(\mathbf{H}) = \infty, \tag{6.26}$$

$$\lim_{\theta_\mathrm{F} \to \frac{\pi}{2}} \mathrm{oSNR}(\mathbf{H}) = 0. \tag{6.27}$$

In other words, $\varrho\left(\mathbf{x}_\mathrm{F}, \mathbf{v}_\mathrm{F}\right)$ is purely real (resp. imaginary) if and only if the output SNR is equal to infinity (resp. zero). In this interpretation, we represent the output SNR by the angle θ_F for which it takes its values between 0 and $\pi/2$.

It is easy to check that $\mathrm{oSNR}(\mathbf{H}) > \mathrm{iSNR}$ is equivalent to $\rho^2\left(\mathbf{x}_\mathrm{F}, \mathbf{z}\right) > \rho^2\left(\mathbf{x}, \mathbf{y}\right)$. In this case, $\rho^2\left(\mathbf{v}_\mathrm{F}, \mathbf{z}\right) < \rho^2\left(\mathbf{v}, \mathbf{y}\right)$. With the angle interpretation, $\mathrm{oSNR}(\mathbf{H}) > \mathrm{iSNR}$ is equivalent to $\cos^2\theta_\mathrm{F} > \cos^2\theta$ which implies that $\theta_\mathrm{F} < \theta$.

To finish this part, we link the noise-reduction and speech-reduction factors with the SPCCs:

$$\frac{\rho^2\left(\mathbf{x}_\mathrm{F}, \mathbf{z}\right)}{\rho^2\left(\mathbf{x}, \mathbf{y}\right)} \cdot \xi_\mathrm{sr}\left(\mathbf{H}\right) = \frac{\rho^2\left(\mathbf{v}_\mathrm{F}, \mathbf{z}\right)}{\rho^2\left(\mathbf{v}, \mathbf{y}\right)} \cdot \xi_\mathrm{nr}\left(\mathbf{H}\right), \tag{6.28}$$

and with the angles:

$$\tan^2\theta \cdot \xi_\mathrm{sr}\left(\mathbf{H}\right) = \tan^2\theta_\mathrm{F} \cdot \xi_\mathrm{nr}\left(\mathbf{H}\right). \tag{6.29}$$

46   6 Fundamental Properties

6.2 In the Frequency Domain

The subband SPCC (or MSCF) between the two signals $X(j\omega)$ and $Y(j\omega)$ [as defined in (2.8)] at frequency ω is

$$\left|\rho\left[X(j\omega), Y(j\omega)\right]\right|^2 = \frac{\phi_x(\omega)}{\phi_y(\omega)}$$

$$= \frac{\text{iSNR}(\omega)}{1 + \text{iSNR}(\omega)}. \tag{6.30}$$

This subband SPCC tells us how much the observation signal is noisy at frequency ω. A value of $\left|\rho\left[X(j\omega), Y(j\omega)\right]\right|^2$ close to 1 implies that the speech is largely dominant while a value of $\left|\rho\left[X(j\omega), Y(j\omega)\right]\right|^2$ close to 0 implies that the noise is largely dominant at frequency ω.

The fullband SPCC between the two signals $X(j\omega)$ and $Y(j\omega)$ is

$$\left|\rho\left(X, Y\right)\right|^2 = \frac{\int_{-\pi}^{\pi} \phi_x(\omega)d\omega}{\int_{-\pi}^{\pi} \phi_y(\omega)d\omega}$$

$$= \frac{\text{iSNR}}{1 + \text{iSNR}}$$

$$= \rho^2\left(\mathbf{x}, \mathbf{y}\right). \tag{6.31}$$

As expected, the SPCC between the two vectors $\mathbf{x}(k)$ and $\mathbf{y}(k)$ is identical to the fullband SPCC between $X(j\omega)$ and $Y(j\omega)$.

The subband SPCC between the two signals $X(j\omega)$ and $Z(j\omega)$ [as defined in (2.13)] at frequency ω is

$$\left|\rho\left[X(j\omega), Z(j\omega)\right]\right|^2 = \left|\rho\left[X(j\omega), Y(j\omega)\right]\right|^2. \tag{6.32}$$

The subband SPCC between $X(j\omega)$ and $Z(j\omega)$ is equal to the subband SPCC between $X(j\omega)$ and $Y(j\omega)$ and does not depend on $H(j\omega)$; the same way the subband input SNR is equal to the subband output SNR and does not depend on $H(j\omega)$.

The fullband SPCC between the two signals $X(j\omega)$ and $Z(j\omega)$ is

$$\left|\rho\left(X, Z\right)\right|^2 = \frac{\left|\int_{-\pi}^{\pi} H(j\omega)\phi_x(\omega)d\omega\right|^2}{\left[\int_{-\pi}^{\pi} \phi_x(\omega)d\omega\right]\left[\int_{-\pi}^{\pi} |H(j\omega)|^2 \phi_y(\omega)d\omega\right]} \tag{6.33}$$

$$= \frac{\left|\int_{-\pi}^{\pi} H(j\omega)\phi_x(\omega)d\omega\right|^2}{\left[\int_{-\pi}^{\pi} \phi_x(\omega)d\omega\right]\left[\int_{-\pi}^{\pi} |H(j\omega)|^2 \phi_x(\omega)d\omega\right]} \cdot \frac{\text{oSNR}(H)}{1 + \text{oSNR}(H)}.$$

Property 6.11. We have

$$\left|\rho\left(X, Z\right)\right|^2 = \left|\rho\left(X, X_{\text{F}}\right)\right|^2 \cdot \left|\rho\left(X_{\text{F}}, Z\right)\right|^2, \tag{6.34}$$

where

$$|\rho(X, X_\mathrm{F})|^2 = \frac{\left|\int_{-\pi}^{\pi} H(j\omega)\phi_x(\omega)d\omega\right|^2}{\left[\int_{-\pi}^{\pi} \phi_x(\omega)d\omega\right]\left[\int_{-\pi}^{\pi} |H(j\omega)|^2 \phi_x(\omega)d\omega\right]} \tag{6.35}$$

and

$$|\rho(X_\mathrm{F}, Z)|^2 = \frac{\mathrm{oSNR}(H)}{1 + \mathrm{oSNR}(H)}. \tag{6.36}$$

The fullband SPCC $|\rho(X, X_\mathrm{F})|^2$ is a speech-distortion index. If $H(j\omega) = 1$, $\forall\omega$ (no speech distortion) then $|\rho(X, X_\mathrm{F})|^2 = 1$. The closer the value of $|\rho(X, X_\mathrm{F})|^2$ is to 0, the more the speech signal is distorted. The fullband SPCC $|\rho(X_\mathrm{F}, Z)|^2$ shows the SNR improvement and reaches its maximum when $\mathrm{oSNR}(H)$ is maximized. The maximization of $|\rho(X, Z)|^2$ leads to an optimal filter.

Property 6.12. We have

$$|\rho(X, Z)|^2 \leq \frac{\mathrm{oSNR}(H)}{1 + \mathrm{oSNR}(H)}, \tag{6.37}$$

with equality when $H(j\omega) = 1$, $\forall\omega$.

Proof. This property follows immediately from (6.34) since $|\rho(X, X_\mathrm{F})|^2 \leq 1$.

Property 6.13. We have

$$|\rho(X_\mathrm{F}, Y)|^2 = |\rho(X, X_\mathrm{F})|^2 \cdot |\rho(X, Y)|^2. \tag{6.38}$$

Proof. Indeed

$$\begin{aligned}
|\rho(X_\mathrm{F}, Y)|^2 &= \frac{\left|\int_{-\pi}^{\pi} H(j\omega)\phi_x(\omega)d\omega\right|^2}{\left[\int_{-\pi}^{\pi} |H(j\omega)|^2 \phi_x(\omega)d\omega\right]\left[\int_{-\pi}^{\pi} \phi_y(\omega)d\omega\right]} \\
&= \frac{\left|\int_{-\pi}^{\pi} H(j\omega)\phi_x(\omega)d\omega\right|^2}{\left[\int_{-\pi}^{\pi} |H(j\omega)|^2 \phi_x(\omega)d\omega\right]\left[\int_{-\pi}^{\pi} \phi_x(\omega)d\omega\right]} \cdot \frac{\mathrm{iSNR}}{1 + \mathrm{iSNR}} \\
&= |\rho(X, X_\mathrm{F})|^2 \cdot |\rho(X, Y)|^2.
\end{aligned}$$

Property 6.14. We have

$$|\rho(X_\mathrm{F}, Y)|^2 \leq \frac{\mathrm{iSNR}}{1 + \mathrm{iSNR}}, \tag{6.39}$$

with equality when $H(j\omega) = 1$, $\forall\omega$.

48   6 Fundamental Properties

Proof. This property follows immediately from (6.38) since $|\rho(X, X_{\mathrm{F}})|^2 \leq 1$.

The MSCF between $V(j\omega)$ and $Y(j\omega)$ is another way to see how much the observation signal is affected by the noise at frequency ω. This MSCF is

$$|\rho[V(j\omega), Y(j\omega)]|^2 = \frac{\phi_v(\omega)}{\phi_y(\omega)}$$

$$= \frac{1}{1 + \mathrm{iSNR}(\omega)}. \qquad (6.40)$$

Property 6.15. We have

$$|\rho[X(j\omega), Y(j\omega)]|^2 + |\rho[V(j\omega), Y(j\omega)]|^2 = 1 \qquad (6.41)$$

and

$$\mathrm{iSNR}(\omega) = \frac{|\rho[X(j\omega), Y(j\omega)]|^2}{|\rho[V(j\omega), Y(j\omega)]|^2}. \qquad (6.42)$$

Proof. Easy to see from (6.30) and (6.40).

The subband input SNR at frequency ω is the ratio of two MSCFs and the sum of these two MSCFs is equal to 1.

The fullband SPCC between $V(j\omega)$ and $Y(j\omega)$ is

$$|\rho(V, Y)|^2 = \frac{1}{1 + \mathrm{iSNR}}$$

$$= \rho^2(\mathbf{v}, \mathbf{y}). \qquad (6.43)$$

The SPCC between the two vectors $\mathbf{v}(k)$ and $\mathbf{y}(k)$ is equal to the fullband SPCC between $V(j\omega)$ and $Y(j\omega)$.

The subband SPCC between the two signals $V(j\omega)$ and $Z(j\omega)$ at frequency ω is

$$|\rho[V(j\omega), Z(j\omega)]|^2 = |\rho[V(j\omega), Y(j\omega)]|^2. \qquad (6.44)$$

The fullband SPCC between the same signals is

$$|\rho(V, Z)|^2 = \frac{\left|\int_{-\pi}^{\pi} H(j\omega)\phi_v(\omega)d\omega\right|^2}{\left[\int_{-\pi}^{\pi} \phi_v(\omega)d\omega\right]\left[\int_{-\pi}^{\pi} |H(j\omega)|^2 \phi_y(\omega)d\omega\right]} \qquad (6.45)$$

$$= \frac{\left|\int_{-\pi}^{\pi} H(j\omega)\phi_v(\omega)d\omega\right|^2}{\left[\int_{-\pi}^{\pi} \phi_v(\omega)d\omega\right]\left[\int_{-\pi}^{\pi} |H(j\omega)|^2 \phi_v(\omega)d\omega\right]} \cdot \frac{1}{1 + \mathrm{oSNR}(H)}.$$

Property 6.16. We have

$$|\rho(V,Z)|^2 = |\rho(V,V_F)|^2 \cdot |\rho(V_F,Z)|^2, \qquad (6.46)$$

where

$$|\rho(V,V_F)|^2 = \frac{\left|\int_{-\pi}^{\pi} H(j\omega)\phi_v(\omega)d\omega\right|^2}{\left[\int_{-\pi}^{\pi} \phi_v(\omega)d\omega\right]\left[\int_{-\pi}^{\pi} |H(j\omega)|^2 \phi_v(\omega)d\omega\right]} \qquad (6.47)$$

and

$$|\rho(V_F,Z)|^2 = \frac{1}{1 + \text{oSNR}(H)}. \qquad (6.48)$$

Property 6.17. We have

$$|\rho(V,Z)|^2 \leq \frac{1}{1 + \text{oSNR}(H)}, \qquad (6.49)$$

with equality when $H(j\omega) = 1$, $\forall \omega$.

Proof. This property follows immediately from (6.46) since $|\rho(V,V_F)|^2 \leq 1$.

Property 6.18. We have

$$|\rho(V_F,Y)|^2 = |\rho(V,V_F)|^2 \cdot |\rho(V,Y)|^2. \qquad (6.50)$$

Proof. Indeed

$$
\begin{aligned}
|\rho(V_F,Y)|^2 &= \frac{\left|\int_{-\pi}^{\pi} H(j\omega)\phi_v(\omega)d\omega\right|^2}{\left[\int_{-\pi}^{\pi} |H(j\omega)|^2 \phi_v(\omega)d\omega\right]\left[\int_{-\pi}^{\pi} \phi_y(\omega)d\omega\right]} \\
&= \frac{\left|\int_{-\pi}^{\pi} H(j\omega)\phi_v(\omega)d\omega\right|^2}{\left[\int_{-\pi}^{\pi} |H(j\omega)|^2 \phi_v(\omega)d\omega\right]\left[\int_{-\pi}^{\pi} \phi_v(\omega)d\omega\right]} \cdot \frac{1}{1 + \text{iSNR}} \\
&= |\rho(V,V_F)|^2 \cdot |\rho(V,Y)|^2.
\end{aligned}
$$

Property 6.19. We have

$$|\rho(V_F,Y)|^2 \leq \frac{1}{1 + \text{iSNR}}, \qquad (6.51)$$

with equality when $H(j\omega) = 1$, $\forall \omega$.

Proof. This property follows immediately from (6.50) since $|\rho(V,V_F)|^2 \leq 1$.

50   6 Fundamental Properties

Property 6.20. We have

$$|\rho(X_\mathrm{F}, Z)|^2 + |\rho(V_\mathrm{F}, Z)|^2 = 1 \tag{6.52}$$

and

$$\mathrm{oSNR}(H) = \frac{|\rho(X_\mathrm{F}, Z)|^2}{|\rho(V_\mathrm{F}, Z)|^2}. \tag{6.53}$$

Proof. Easy to see from (6.36) and (6.48).

The fullband output SNR for the frequency-domain approach is the ratio of two fullband SPCCs and the sum of these two SPCCs is equal to 1.

It can be checked that $\mathrm{oSNR}(H) > \mathrm{iSNR}$ is equivalent to $|\rho(X_\mathrm{F}, Z)|^2 > \rho^2(\mathbf{x}, \mathbf{y})$. In this situation, $|\rho(V_\mathrm{F}, Z)|^2 < \rho^2(\mathbf{v}, \mathbf{y})$.

6.3 In the KLE Domain

The subband SPCC (or MSCF) between the two signals $c_{x,l}(k)$ and $c_{y,l}(k)$ [as defined in (2.22)] in the subband l is

$$\begin{aligned}
\rho^2(c_{x,l}, c_{y,l}) &= \frac{\lambda_l}{\lambda_l + \mathbf{q}_l^T \mathbf{R}_v \mathbf{q}_l} \\
&= \frac{\mathrm{iSNR}_l}{1 + \mathrm{iSNR}_l}.
\end{aligned} \tag{6.54}$$

This SPCC tells us how much the observation signal is noisy in the subband l. A value of $\rho^2(c_{x,l}, c_{y,l})$ close to 1 implies that the speech is largely dominant while a value of $\rho^2(c_{x,l}, c_{y,l})$ close to 0 implies that the noise is largely dominant in the subband l.

The subband SPCC between the two signals $c_{x,l}(k)$ and $c_{z,l}(k)$ [as defined in (2.26)] in the subband l is

$$\begin{aligned}
\rho^2(c_{x,l}, c_{z,l}) &= \frac{\left(\mathbf{i}_l^T \mathbf{R}_{c_x,l} \mathbf{h}_l\right)^2}{\lambda_l \cdot \mathbf{h}_l^T \mathbf{R}_{c_y,l} \mathbf{h}_l} \\
&= \frac{\left(\mathbf{i}_l^T \mathbf{R}_{c_x,l} \mathbf{h}_l\right)^2}{\lambda_l \cdot \mathbf{h}_l^T \mathbf{R}_{c_x,l} \mathbf{h}_l} \cdot \frac{\mathrm{oSNR}(\mathbf{h}_l)}{1 + \mathrm{oSNR}(\mathbf{h}_l)}.
\end{aligned} \tag{6.55}$$

Property 6.21. We have

$$\rho^2(c_{x,l}, c_{z,l}) = \rho^2\left(c_{x,l}, \mathbf{h}_l^T \mathbf{c}_{x,l}\right) \cdot \rho^2\left(\mathbf{h}_l^T \mathbf{c}_{x,l}, c_{z,l}\right), \tag{6.56}$$

where

$$\rho^2\left(c_{x,l}, \mathbf{h}_l^T \mathbf{c}_{x,l}\right) = \frac{\left(\mathbf{i}_l^T \mathbf{R}_{c_x,l} \mathbf{h}_l\right)^2}{\lambda_l \cdot \mathbf{h}_l^T \mathbf{R}_{c_x,l} \mathbf{h}_l} \tag{6.57}$$

and

$$\rho^2\left(\mathbf{h}_l^T \mathbf{c}_{x,l}, c_{z,l}\right) = \frac{\text{oSNR}(\mathbf{h}_l)}{1 + \text{oSNR}(\mathbf{h}_l)}. \tag{6.58}$$

Property 6.22. We have

$$\rho^2\left(c_{x,l}, c_{z,l}\right) \leq \frac{\text{oSNR}(\mathbf{h}_l)}{1 + \text{oSNR}(\mathbf{h}_l)}, \tag{6.59}$$

with equality when $\mathbf{h}_l = \mathbf{i}_l$.

Proof. This property follows immediately from (6.56) since $\rho^2\left(c_{x,l}, \mathbf{h}_l^T \mathbf{c}_{x,l}\right)$.

Property 6.23. We have

$$\rho^2\left(\mathbf{h}_l^T \mathbf{c}_{x,l}, c_{y,l}\right) = \rho^2\left(c_{x,l}, \mathbf{h}_l^T \mathbf{c}_{x,l}\right) \cdot \rho^2\left(c_{x,l}, c_{y,l}\right). \tag{6.60}$$

Proof. Indeed

$$\begin{aligned}
\rho^2\left(\mathbf{h}_l^T \mathbf{c}_{x,l}, c_{y,l}\right) &= \frac{\left(\mathbf{i}_l^T \mathbf{R}_{c_x,l} \mathbf{h}_l\right)^2}{\mathbf{h}_l^T \mathbf{R}_{c_x,l} \mathbf{h}_l \cdot \left(\lambda_l + \mathbf{q}_l^T \mathbf{R}_v \mathbf{q}_l\right)} \\
&= \frac{\left(\mathbf{i}_l^T \mathbf{R}_{c_x,l} \mathbf{h}_l\right)^2}{\mathbf{h}_l^T \mathbf{R}_{c_x,l} \mathbf{h}_l \cdot \lambda_l} \cdot \frac{\text{iSNR}_l}{1 + \text{iSNR}_l} \\
&= \rho^2\left(c_{x,l}, \mathbf{h}_l^T \mathbf{c}_{x,l}\right) \cdot \rho^2\left(c_{x,l}, c_{y,l}\right).
\end{aligned}$$

Property 6.24. We have

$$\rho^2\left(\mathbf{h}_l^T \mathbf{c}_{x,l}, c_{y,l}\right) \leq \frac{\text{iSNR}_l}{1 + \text{iSNR}_l}, \tag{6.61}$$

with equality when $\mathbf{h}_l = \mathbf{i}_l$.

Proof. This property follows immediately from (6.60) since $\rho^2\left(c_{x,l}, \mathbf{h}_l^T \mathbf{c}_{x,l}\right) \leq 1$.

The SPCC between the two signals $c_{v,l}(k)$ and $c_{y,l}(k)$ is another way to see how much the microphone signal is affected by the noise in the subband l. This SPCC is

$$\begin{aligned}
\rho^2\left(c_{v,l}, c_{y,l}\right) &= \frac{\mathbf{q}_l^T \mathbf{R}_v \mathbf{q}_l}{\mathbf{q}_l^T \mathbf{R}_y \mathbf{q}_l} \\
&= \frac{1}{1 + \text{iSNR}_l}. \tag{6.62}
\end{aligned}$$

So a value of $\rho^2(c_{v,l}, c_{y,l})$ close to 1 implies that the noise is largely dominant while a value of $\rho^2(c_{v,l}, c_{y,l})$ close to 0 implies that the speech is largely dominant in the subband l.

Property 6.25. We have

$$\rho^2(c_{x,l}, c_{y,l}) + \rho^2(c_{v,l}, c_{y,l}) = 1 \tag{6.63}$$

and

$$\text{iSNR}_l = \frac{\rho^2(c_{x,l}, c_{y,l})}{\rho^2(c_{v,l}, c_{y,l})}. \tag{6.64}$$

Proof. Easy to see from (6.54) and (6.62).

The subband input SNR is the ratio between two SPCCs and the sum of these two SPCCs is equal to 1.

The SPCC between the two signals $c_{v,l}(k)$ and $c_{z,l}(k)$ in the subband l is

$$\begin{aligned}
\rho^2(c_{v,l}, c_{z,l}) &= \frac{\left(\mathbf{i}_l^T \mathbf{R}_{c_v,l} \mathbf{h}_l\right)^2}{\mathbf{q}_l^T \mathbf{R}_v \mathbf{q}_l \cdot \mathbf{h}_l^T \mathbf{R}_{c_y,l} \mathbf{h}_l} \\
&= \frac{\left(\mathbf{i}_l^T \mathbf{R}_{c_v,l} \mathbf{h}_l\right)^2}{\mathbf{q}_l^T \mathbf{R}_v \mathbf{q}_l \cdot \mathbf{h}_l^T \mathbf{R}_{c_v,l} \mathbf{h}_l} \cdot \frac{1}{1 + \text{oSNR}(\mathbf{h}_l)}.
\end{aligned} \tag{6.65}$$

Property 6.26. We have

$$\rho^2(c_{v,l}, c_{z,l}) = \rho^2\left(c_{v,l}, \mathbf{h}_l^T \mathbf{c}_{v,l}\right) \cdot \rho^2\left(\mathbf{h}_l^T \mathbf{c}_{v,l}, c_{z,l}\right), \tag{6.66}$$

where

$$\rho^2\left(c_{v,l}, \mathbf{h}_l^T \mathbf{c}_{v,l}\right) = \frac{\left(\mathbf{i}_l^T \mathbf{R}_{c_v,l} \mathbf{h}_l\right)^2}{\mathbf{q}_l^T \mathbf{R}_v \mathbf{q}_l \cdot \mathbf{h}_l^T \mathbf{R}_{c_v,l} \mathbf{h}_l} \tag{6.67}$$

and

$$\rho^2\left(\mathbf{h}_l^T \mathbf{c}_{v,l}, c_{z,l}\right) = \frac{1}{1 + \text{oSNR}(\mathbf{h}_l)}. \tag{6.68}$$

Property 6.27. We have

$$\rho^2(c_{v,l}, c_{z,l}) \leq \frac{1}{1 + \text{oSNR}(\mathbf{h}_l)}, \tag{6.69}$$

with equality when $\mathbf{h}_l = \mathbf{i}_l$.

Proof. This property follows immediately from (6.66) since $\rho^2\left(c_{v,l}, \mathbf{h}_l^T \mathbf{c}_{v,l}\right) \leq 1$.

Property 6.28. We have

$$\rho^2 \left(\mathbf{h}_l^T \mathbf{c}_{v,l}, c_{y,l} \right) = \rho^2 \left(c_{v,l}, \mathbf{h}_l^T \mathbf{c}_{v,l} \right) \cdot \rho^2 \left(c_{v,l}, c_{y,l} \right). \tag{6.70}$$

Proof. Indeed

$$\begin{aligned}
\rho^2 \left(\mathbf{h}_l^T \mathbf{c}_{v,l}, c_{y,l} \right) &= \frac{\left(\mathbf{i}_l^T \mathbf{R}_{c_v,l} \mathbf{h}_l \right)^2}{\mathbf{h}_l^T \mathbf{R}_{c_v,l} \mathbf{h}_l \cdot \left(\lambda_l + \mathbf{q}_l^T \mathbf{R}_v \mathbf{q}_l \right)} \\
&= \frac{\left(\mathbf{i}_l^T \mathbf{R}_{c_v,l} \mathbf{h}_l \right)^2}{\mathbf{h}_l^T \mathbf{R}_{c_v,l} \mathbf{h}_l \cdot \mathbf{q}_l^T \mathbf{R}_v \mathbf{q}_l} \cdot \frac{1}{1 + \mathrm{iSNR}_l} \\
&= \rho^2 \left(c_{v,l}, \mathbf{h}_l^T \mathbf{c}_{v,l} \right) \cdot \rho^2 \left(c_{v,l}, c_{y,l} \right).
\end{aligned}$$

Property 6.29. We have

$$\rho^2 \left(\mathbf{h}_l^T \mathbf{c}_{v,l}, c_{y,l} \right) \le \frac{1}{1 + \mathrm{iSNR}_l}, \tag{6.71}$$

with equality when $\mathbf{h}_l = \mathbf{i}_l$.

Proof. This property follows immediately from (6.70) since $\rho^2 \left(c_{v,l}, \mathbf{h}_l^T \mathbf{c}_{v,l} \right) \le 1$.

Property 6.30. We have

$$\rho^2 \left(\mathbf{h}_l^T \mathbf{c}_{x,l}, c_{z,l} \right) + \rho^2 \left(\mathbf{h}_l^T \mathbf{c}_{v,l}, c_{z,l} \right) = 1 \tag{6.72}$$

and

$$\mathrm{oSNR}(\mathbf{h}_l) = \frac{\rho^2 \left(\mathbf{h}_l^T \mathbf{c}_{x,l}, c_{z,l} \right)}{\rho^2 \left(\mathbf{h}_l^T \mathbf{c}_{v,l}, c_{z,l} \right)}. \tag{6.73}$$

Proof. Easy to see from (6.58) and (6.68).

The subband output SNR for the KLE-domain approach is the ratio of two SPCCs depending on $\mathbf{h}_l$ and the sum of these two SPCCs is equal to 1.

For $L_l > 1$, it is easy to see that $\mathrm{oSNR}(\mathbf{h}_l) > \mathrm{iSNR}$ is equivalent to $\rho^2 \left(\mathbf{h}_l^T \mathbf{c}_{x,l}, c_{z,l} \right) > \rho^2 \left(c_{x,l}, c_{y,l} \right)$. In this case, $\rho^2 \left(\mathbf{h}_l^T \mathbf{c}_{v,l}, c_{z,l} \right) < \rho^2 \left(c_{v,l}, c_{y,l} \right)$.
Now, let us define the four vectors of length L:

$$\mathbf{c}_v(k) = \begin{bmatrix} c_{v,1}(k) & c_{v,2}(k) & \cdots & c_{v,L}(k) \end{bmatrix}^T, \tag{6.74a}$$

$$\mathbf{c}_x(k) = \begin{bmatrix} c_{x,1}(k) & c_{x,2}(k) & \cdots & c_{x,L}(k) \end{bmatrix}^T, \tag{6.74b}$$

$$\mathbf{c}_y(k) = \begin{bmatrix} c_{y,1}(k) & c_{y,2}(k) & \cdots & c_{y,L}(k) \end{bmatrix}^T, \tag{6.74c}$$

$$\mathbf{c}_z(k) = \begin{bmatrix} c_{z,1}(k) & c_{z,2}(k) & \cdots & c_{z,L}(k) \end{bmatrix}^T, \tag{6.74d}$$

54   6 Fundamental Properties

and the two filtered vectors of length L:

$$\mathbf{c}_{v,\mathrm{F}}(k) = \left[\mathbf{h}_1^T \mathbf{c}_{v,1}(k) \ \mathbf{h}_2^T \mathbf{c}_{v,2}(k) \ \cdots \ \mathbf{h}_L^T \mathbf{c}_{v,L}(k) \right]^T, \tag{6.75a}$$

$$\mathbf{c}_{x,\mathrm{F}}(k) = \left[\mathbf{h}_1^T \mathbf{c}_{x,1}(k) \ \mathbf{h}_2^T \mathbf{c}_{x,2}(k) \ \cdots \ \mathbf{h}_L^T \mathbf{c}_{x,L}(k) \right]^T. \tag{6.75b}$$

The SPCC between the two vectors $\mathbf{c}_x(k)$ and $\mathbf{c}_y(k)$ is

$$
\begin{aligned}
\rho^2\left(\mathbf{c}_x, \mathbf{c}_y\right) &= \frac{\sum_{l=1}^{L} \lambda_l}{\sum_{l=1}^{L} \lambda_l + \sum_{l}^{L} \mathbf{q}_l^T \mathbf{R}_v \mathbf{q}_l} \\
&= \frac{\mathrm{iSNR}}{1 + \mathrm{iSNR}} \\
&= \rho^2\left(\mathbf{x}, \mathbf{y}\right).
\end{aligned}
\tag{6.76}
$$

As expected, the SPCC between the two vectors $\mathbf{x}(k)$ and $\mathbf{y}(k)$ is identical to the SPCC between the two vectors $\mathbf{c}_x(k)$ and $\mathbf{c}_y(k)$.

The SPCC between the two vectors $\mathbf{c}_x(k)$ and $\mathbf{c}_z(k)$ is

$$
\begin{aligned}
\rho^2\left(\mathbf{c}_x, \mathbf{c}_z\right) &= \frac{\left(\sum_{l=1}^{L} \mathbf{i}_l^T \mathbf{R}_{c_x,l} \mathbf{h}_l\right)^2}{\sum_{l=1}^{L} \lambda_l \cdot \sum_{l=1}^{L} \mathbf{h}_l^T \mathbf{R}_{c_y,l} \mathbf{h}_l} \\
&= \frac{\left(\sum_{l=1}^{L} \mathbf{i}_l^T \mathbf{R}_{c_x,l} \mathbf{h}_l\right)^2}{\sum_{l=1}^{L} \lambda_l \cdot \sum_{l=1}^{L} \mathbf{h}_l^T \mathbf{R}_{c_x,l} \mathbf{h}_l} \cdot \frac{\mathrm{oSNR}(\mathbf{h}_{1:L})}{1 + \mathrm{oSNR}(\mathbf{h}_{1:L})}.
\end{aligned}
\tag{6.77}
$$

Property 6.31. We have

$$\rho^2\left(\mathbf{c}_x, \mathbf{c}_z\right) = \rho^2\left(\mathbf{c}_x, \mathbf{c}_{x,\mathrm{F}}\right) \cdot \rho^2\left(\mathbf{c}_{x,\mathrm{F}}, \mathbf{c}_z\right), \tag{6.78}$$

where

$$\rho^2\left(\mathbf{c}_x, \mathbf{c}_{x,\mathrm{F}}\right) = \frac{\left(\sum_{l=1}^{L} \mathbf{i}_l^T \mathbf{R}_{c_x,l} \mathbf{h}_l\right)^2}{\sum_{l=1}^{L} \lambda_l \cdot \sum_{l=1}^{L} \mathbf{h}_l^T \mathbf{R}_{c_x,l} \mathbf{h}_l} \tag{6.79}$$

and

$$\rho^2\left(\mathbf{c}_{x,\mathrm{F}}, \mathbf{c}_z\right) = \frac{\mathrm{oSNR}(\mathbf{h}_{1:L})}{1 + \mathrm{oSNR}(\mathbf{h}_{1:L})}. \tag{6.80}$$

Property 6.32. We have

$$\rho^2\left(\mathbf{c}_x, \mathbf{c}_z\right) \leq \frac{\mathrm{oSNR}(\mathbf{h}_{1:L})}{1 + \mathrm{oSNR}(\mathbf{h}_{1:L})}, \tag{6.81}$$

with equality when $\mathbf{h}_l = \mathbf{i}_l$, $\forall l$.

Proof. This property follows immediately from (6.78) since $\rho^2\left(\mathbf{c}_x, \mathbf{c}_{x,\mathrm{F}}\right) \leq 1$.

Property 6.33. We have

$$\rho^2\left(\mathbf{c}_{x,\mathrm{F}}, \mathbf{c}_y\right) = \rho^2\left(\mathbf{c}_x, \mathbf{c}_{x,\mathrm{F}}\right) \cdot \rho^2\left(\mathbf{c}_x, \mathbf{c}_y\right). \tag{6.82}$$

Proof. Indeed

$$\begin{aligned}
\rho^2\left(\mathbf{c}_{x,\mathrm{F}}, \mathbf{c}_y\right) &= \frac{\left(\sum_{l=1}^{L} \mathbf{i}_l^T \mathbf{R}_{c_x,l} \mathbf{h}_l\right)^2}{\sum_{l=1}^{L} \mathbf{h}_l^T \mathbf{R}_{c_x,l} \mathbf{h}_l \cdot \sum_{l=1}^{L} \left(\lambda_l + \mathbf{q}_l^T \mathbf{R}_v \mathbf{q}_l\right)} \\
&= \frac{\left(\sum_{l=1}^{L} \mathbf{i}_l^T \mathbf{R}_{c_x,l} \mathbf{h}_l\right)^2}{\sum_{l=1}^{L} \mathbf{h}_l^T \mathbf{R}_{c_x,l} \mathbf{h}_l \cdot \sum_{l=1}^{L} \lambda_l} \cdot \frac{\mathrm{iSNR}}{1 + \mathrm{iSNR}} \\
&= \rho^2\left(\mathbf{c}_x, \mathbf{c}_{x,\mathrm{F}}\right) \cdot \rho^2\left(\mathbf{c}_x, \mathbf{c}_y\right).
\end{aligned}$$

Property 6.34. We have

$$\rho^2\left(\mathbf{c}_{x,\mathrm{F}}, \mathbf{c}_y\right) \le \frac{\mathrm{iSNR}}{1 + \mathrm{iSNR}}, \tag{6.83}$$

with equality when $\mathbf{h}_l = \mathbf{i}_l, \ \forall l$.

Proof. This property follows immediately from (6.82) since $\rho^2\left(\mathbf{c}_x, \mathbf{c}_{x,\mathrm{F}}\right) \le 1$.

The SPCC between the two vectors $\mathbf{c}_v$ and $\mathbf{c}_y$ is

$$\begin{aligned}
\rho^2\left(\mathbf{c}_v, \mathbf{c}_y\right) &= \frac{\sum_{l=1}^{L} \mathbf{q}_l^T \mathbf{R}_v \mathbf{q}_l}{\sum_{l=1}^{L} \mathbf{q}_l^T \mathbf{R}_y \mathbf{q}_l} \\
&= \frac{1}{1 + \mathrm{iSNR}} \\
&= \rho^2\left(\mathbf{v}, \mathbf{y}\right). \tag{6.84}
\end{aligned}$$

As expected, the SPCC between the two vectors $\mathbf{v}(k)$ and $\mathbf{y}(k)$ is identical to the SPCC between the two vectors $\mathbf{c}_v(k)$ and $\mathbf{c}_y(k)$.

The SPCC between the two vectors $\mathbf{c}_v(k)$ and $\mathbf{c}_z(k)$ is

$$\begin{aligned}
\rho^2\left(\mathbf{c}_v, \mathbf{c}_z\right) &= \frac{\left(\sum_{l=1}^{L} \mathbf{i}_l^T \mathbf{R}_{c_v,l} \mathbf{h}_l\right)^2}{\sum_{l=1}^{L} \mathbf{q}_l^T \mathbf{R}_v \mathbf{q}_l \cdot \sum_{l=1}^{L} \mathbf{h}_l^T \mathbf{R}_{c_y,l} \mathbf{h}_l} \\
&= \frac{\left(\sum_{l=1}^{L} \mathbf{i}_l^T \mathbf{R}_{c_v,l} \mathbf{h}_l\right)^2}{\sum_{l=1}^{L} \mathbf{q}_l^T \mathbf{R}_v \mathbf{q}_l \cdot \sum_{l=1}^{L} \mathbf{h}_l^T \mathbf{R}_{c_v,l} \mathbf{h}_l} \cdot \frac{1}{1 + \mathrm{oSNR}(\mathbf{h}_{1:L})}. \tag{6.85}
\end{aligned}$$

Property 6.35. We have

$$\rho^2\left(\mathbf{c}_v, \mathbf{c}_z\right) = \rho^2\left(\mathbf{c}_v, \mathbf{c}_{v,\mathrm{F}}\right) \cdot \rho^2\left(\mathbf{c}_{v,\mathrm{F}}, \mathbf{c}_z\right), \tag{6.86}$$

where

56   6 Fundamental Properties

$$\rho^2\left(\mathbf{c}_v, \mathbf{c}_{v,\mathrm{F}}\right) = \frac{\left(\sum_{l=1}^{L} \mathbf{i}_l^T \mathbf{R}_{c_v,l} \mathbf{h}_l\right)^2}{\sum_{l=1}^{L} \mathbf{q}_l^T \mathbf{R}_v \mathbf{q}_l \cdot \sum_{l=1}^{L} \mathbf{h}_l^T \mathbf{R}_{c_v,l} \mathbf{h}_l} \qquad (6.87)$$

and

$$\rho^2\left(\mathbf{c}_{v,\mathrm{F}}, \mathbf{c}_z\right) = \frac{1}{1 + \mathrm{oSNR}(\mathbf{h}_{1:L})}. \qquad (6.88)$$

Property 6.36. We have

$$\rho^2\left(\mathbf{c}_v, \mathbf{c}_z\right) \leq \frac{1}{1 + \mathrm{oSNR}(\mathbf{h}_{1:L})}, \qquad (6.89)$$

with equality when $\mathbf{h}_l = \mathbf{i}_l$, $\forall l$.

Proof. This property follows immediately from (6.86) since $\rho^2\left(\mathbf{c}_v, \mathbf{c}_{v,\mathrm{F}}\right) \leq 1$.

Property 6.37. We have

$$\rho^2\left(\mathbf{c}_{v,\mathrm{F}}, \mathbf{c}_y\right) = \rho^2\left(\mathbf{c}_v, \mathbf{c}_{v,\mathrm{F}}\right) \cdot \rho^2\left(\mathbf{c}_v, \mathbf{c}_y\right). \qquad (6.90)$$

Proof. Indeed

$$\begin{aligned}
\rho^2\left(\mathbf{c}_{v,\mathrm{F}}, \mathbf{c}_y\right) &= \frac{\left(\sum_{l=1}^{L} \mathbf{i}_l^T \mathbf{R}_{c_v,l} \mathbf{h}_l\right)^2}{\sum_{l=1}^{L} \mathbf{h}_l^T \mathbf{R}_{c_v,l} \mathbf{h}_l \cdot \sum_{l=1}^{L} \left(\lambda_l + \mathbf{q}_l^T \mathbf{R}_v \mathbf{q}_l\right)} \\
&= \frac{\left(\sum_{l=1}^{L} \mathbf{i}_l^T \mathbf{R}_{c_v,l} \mathbf{h}_l\right)^2}{\sum_{l=1}^{L} \mathbf{h}_l^T \mathbf{R}_{c_v,l} \mathbf{h}_l \cdot \sum_{l=1}^{L} \mathbf{q}_l^T \mathbf{R}_v \mathbf{q}_l} \cdot \frac{1}{1 + \mathrm{iSNR}} \\
&= \rho^2\left(\mathbf{c}_v, \mathbf{c}_{v,\mathrm{F}}\right) \cdot \rho^2\left(\mathbf{c}_v, \mathbf{c}_y\right).
\end{aligned}$$

Property 6.38. We have

$$\rho^2\left(\mathbf{c}_{v,\mathrm{F}}, \mathbf{c}_y\right) \leq \frac{1}{1 + \mathrm{iSNR}}, \qquad (6.91)$$

with equality when $\mathbf{h}_l = \mathbf{i}_l$, $\forall l$.

Proof. This property follows immediately from (6.90) since $\rho^2\left(\mathbf{c}_v, \mathbf{c}_{v,\mathrm{F}}\right) \leq 1$.

Property 6.39. We have

$$\rho^2\left(\mathbf{c}_{x,\mathrm{F}}, \mathbf{c}_z\right) + \rho^2\left(\mathbf{c}_{v,\mathrm{F}}, \mathbf{c}_z\right) = 1 \qquad (6.92)$$

and

$$\mathrm{oSNR}(\mathbf{h}_{1:L}) = \frac{\rho^2\left(\mathbf{c}_{x,\mathrm{F}}, \mathbf{c}_z\right)}{\rho^2\left(\mathbf{c}_{v,\mathrm{F}}, \mathbf{c}_z\right)}. \qquad (6.93)$$

Proof. Easy to see from (6.80) and (6.88).

The fullband output SNR for the KLE-domain approach is the ratio of two SPCCs depending on $\mathbf{h}_l$, $l = 1, 2, \ldots, L$ and the sum of these two SPCCs is equal to 1.

It can be checked that $\mathrm{oSNR}(\mathbf{h}_{1:L}) > \mathrm{iSNR}$ is equivalent to $\rho^2\left(\mathbf{c}_{x,\mathrm{F}}, \mathbf{c}_z\right) > \rho^2\left(\mathbf{c}_x, \mathbf{c}_y\right)$. In this case, $\rho^2\left(\mathbf{c}_{v,\mathrm{F}}, \mathbf{c}_z\right) < \rho^2\left(\mathbf{c}_v, \mathbf{c}_y\right)$.

6.4 Summary

In this chapter, we discussed many interesting properties of the SPCC in different domains. These fundamental properties establish relationships between the SPCC and the input and output SNRs. Armed with these properties, it is possible in many cases to derive and analyze optimal noise reduction filters by a simple inspection of a certain form of the SPCC, which will be exploited in the next several chapters.

7

Optimal Filters in the Time Domain

Having discussed different error criteria and performance measures, we now begin our search for reliable and practical noise reduction filters. In this chapter, we restrict our attention to filters in the time domain. Particularly, much emphasis is on the Wiener filter [125] as it is directly derived from the MSE criterion and most well-known algorithms are somehow related to it.

7.1 Wiener Filter

If we differentiate the MSE criterion, $J(\mathbf{H})$ [eq. (4.5)], with respect to $\mathbf{H}$ and equate the result to zero, we easily find the Wiener filter matrix:

$$\begin{aligned} \mathbf{H}_{\mathrm{W}}^{T} &= \mathbf{R}_y^{-1}\mathbf{R}_x \\ &= \mathbf{I} - \mathbf{R}_y^{-1}\mathbf{R}_v. \end{aligned} \tag{7.1}$$

This optimal filter depends on the correlation matrices $\mathbf{R}_y$ and $\mathbf{R}_v$: the first one can be estimated during speech-and-noise periods while the second one can be estimated during noise-only intervals assuming that the statistics of the noise do not change much with time.

Now, if we substitute (2.16) into (7.1) we get another useful form of the time-domain Wiener filter:

$$\mathbf{H}_{\mathrm{W}}^{T} = \mathbf{Q}\left[\mathbf{\Lambda} + \mathbf{Q}^{T}\mathbf{R}_v\mathbf{Q}\right]^{-1}\mathbf{\Lambda}\mathbf{Q}^{T}. \tag{7.2}$$

Let us define the following normalized correlation matrices:

$$\tilde{\mathbf{R}}_v = \frac{\mathbf{R}_v}{\sigma_v^2},$$

$$\tilde{\mathbf{R}}_x = \frac{\mathbf{R}_x}{\sigma_x^2},$$

$$\tilde{\mathbf{R}}_y = \frac{\mathbf{R}_y}{\sigma_y^2}.$$

J. Benesty et al., *Noise Reduction in Speech Processing*, Springer Topics in Signal Processing 2,
DOI 10.1007/978-3-642-00296-0_7, © Springer-Verlag Berlin Heidelberg 2009

60   7 Optimal Filters in the Time Domain

A third way to write the Wiener filter is

$$\mathbf{H}_{\mathrm{W}}^{T} = \left[\frac{\mathbf{I}}{\mathrm{iSNR}} + \tilde{\mathbf{R}}_{v}^{-1}\tilde{\mathbf{R}}_{x} \right]^{-1} \tilde{\mathbf{R}}_{v}^{-1}\tilde{\mathbf{R}}_{x} \tag{7.3}$$
$$= \rho^{2}(\mathbf{x},\mathbf{y})\tilde{\mathbf{R}}_{y}^{-1}\tilde{\mathbf{R}}_{x}$$
$$= \mathbf{I} - \rho^{2}(\mathbf{v},\mathbf{y})\tilde{\mathbf{R}}_{y}^{-1}\tilde{\mathbf{R}}_{v},$$

We can see from (7.3) that

$$\lim_{\mathrm{iSNR}\to\infty} \mathbf{H}_{\mathrm{W}} = \mathbf{I}, \tag{7.4}$$
$$\lim_{\mathrm{iSNR}\to 0} \mathbf{H}_{\mathrm{W}} = \mathbf{0}_{L\times L}, \tag{7.5}$$

where $\mathbf{0}_{L\times L}$ is an $L \times L$ matrix with all its elements being zeros. Clearly, the Wiener filter may have a disastrous effect for low input SNRs since it may remove everything (noise and speech).

Property 7.1. With the optimal Wiener filter given in (7.1), the output SNR is always greater than or equal to the input SNR, i.e., oSNR($\mathbf{H}_{\mathrm{W}}$) $\geq$ iSNR.

Proof. Let us evaluate the SPCC between the two vectors $\mathbf{y}(k)$ and $\mathbf{z}_{\mathrm{W}}(k) = \mathbf{H}_{\mathrm{W}}\mathbf{y}(k)$:

$$\rho^{2}(\mathbf{y},\mathbf{z}_{\mathrm{W}}) = \frac{\mathrm{tr}^{2}\left(\mathbf{H}_{\mathrm{W}}\mathbf{R}_{y}\right)}{\mathrm{tr}\left(\mathbf{R}_{y}\right) \cdot \mathrm{tr}\left(\mathbf{H}_{\mathrm{W}}\mathbf{R}_{y}\mathbf{H}_{\mathrm{W}}^{T}\right)}$$
$$= \frac{\mathrm{tr}\left(\mathbf{R}_{x}\right)}{\mathrm{tr}\left(\mathbf{R}_{y}\right)} \cdot \frac{\mathrm{tr}\left(\mathbf{R}_{x}\right)}{\mathrm{tr}\left(\mathbf{H}_{\mathrm{W}}\mathbf{R}_{x}\right)}$$
$$= \frac{\rho^{2}(\mathbf{x},\mathbf{y})}{\rho^{2}(\mathbf{x},\mathbf{z}_{\mathrm{W}})}.$$

Therefore

$$\rho^{2}(\mathbf{x},\mathbf{y}) = \rho^{2}(\mathbf{y},\mathbf{z}_{\mathrm{W}}) \cdot \rho^{2}(\mathbf{x},\mathbf{z}_{\mathrm{W}}) \leq \rho^{2}(\mathbf{x},\mathbf{z}_{\mathrm{W}}).$$

Using (6.1) and Property 6.2 in the previous expression, we get

$$\frac{\mathrm{iSNR}}{1 + \mathrm{iSNR}} \leq \frac{\mathrm{oSNR}(\mathbf{H}_{\mathrm{W}})}{1 + \mathrm{oSNR}(\mathbf{H}_{\mathrm{W}})},$$

that we can slightly rearranged to

$$\frac{1}{1 + \frac{1}{\mathrm{iSNR}}} \leq \frac{1}{1 + \frac{1}{\mathrm{oSNR}(\mathbf{H}_{\mathrm{W}})}},$$

which implies that

$$\frac{1}{\text{iSNR}} \geq \frac{1}{\text{oSNR}(\mathbf{H}_\text{W})}.$$

As a result

$$\text{oSNR}(\mathbf{H}_\text{W}) \geq \text{iSNR}.$$

The minimum MSE (MMSE) and minimum NMSE (MNMSE) are obtained by replacing $\mathbf{H}_\text{W}$ in (4.5) and (4.8):

$$J(\mathbf{H}_\text{W}) = \text{tr}(\mathbf{R}_x) - \text{tr}\left(\mathbf{R}_x \mathbf{R}_y^{-1} \mathbf{R}_x\right)$$
$$= \text{tr}(\mathbf{R}_v) - \text{tr}\left(\mathbf{R}_v \mathbf{R}_y^{-1} \mathbf{R}_v\right), \tag{7.6}$$

$$\tilde{J}(\mathbf{H}_\text{W}) = 1 - \frac{\text{tr}\left(\mathbf{R}_v \mathbf{R}_y^{-1} \mathbf{R}_v\right)}{\text{tr}(\mathbf{R}_v)} \leq 1. \tag{7.7}$$

We can compute the speech-distortion index by substituting (7.1) into (3.23):

$$\upsilon_\text{sd}(\mathbf{H}_\text{W}) = 1 - \frac{\text{oSNR}(\mathbf{H}_\text{W}) + 2}{\text{iSNR} \cdot \xi_\text{nr}(\mathbf{H}_\text{W})} \leq 1. \tag{7.8}$$

Using (4.8) and (7.8), we get the noise-reduction factor:

$$\xi_\text{nr}(\mathbf{H}_\text{W}) = \frac{\text{oSNR}(\mathbf{H}_\text{W}) + 1}{\text{iSNR} - \tilde{J}(\mathbf{H}_\text{W})} \geq 1. \tag{7.9}$$

Property 7.2. We have

$$\tilde{J}(\mathbf{H}_\text{W}) = \text{iSNR}\left[1 - \rho^2(\mathbf{x}, \mathbf{z}_\text{W})\right]. \tag{7.10}$$

Proof. Indeed

$$\tilde{J}(\mathbf{H}_\text{W}) = \frac{\text{tr}(\mathbf{R}_x)}{\text{tr}(\mathbf{R}_v)} - \frac{\text{tr}\left(\mathbf{R}_x \mathbf{R}_y^{-1} \mathbf{R}_x\right)}{\text{tr}(\mathbf{R}_v)}$$
$$= \text{iSNR}\left[1 - \frac{\text{tr}\left(\mathbf{R}_x \mathbf{R}_y^{-1} \mathbf{R}_x\right)}{\text{tr}(\mathbf{R}_x)}\right]$$
$$= \text{iSNR}\left[1 - \rho^2(\mathbf{x}, \mathbf{z}_\text{W})\right].$$

Therefore, the NMSE is minimized when the SPCC between the two vectors $\mathbf{x}(k)$ and $\mathbf{z}(k)$ is maximized. This SPCC can be rewritten as follows:

$$\rho^2(\mathbf{x}, \mathbf{z}_\text{W}) = \frac{1}{\xi_\text{sr}(\mathbf{H}_\text{W})} \cdot \frac{1 + \text{oSNR}(\mathbf{H}_\text{W})}{\text{oSNR}(\mathbf{H}_\text{W})}. \tag{7.11}$$

We observe that the Wiener filter is compromising between speech reduction (i.e., speech distortion) and output SNR improvement.

62   7 Optimal Filters in the Time Domain

Property 7.3. We have

$$\frac{\text{iSNR}}{1 + \text{oSNR}(\mathbf{H}_\text{W})} \leq \tilde{J}(\mathbf{H}_\text{W}) \leq \frac{\text{iSNR}}{1 + \text{iSNR}}. \tag{7.12}$$

Proof. Since

$$\rho^2(\mathbf{x}, \mathbf{z}_\text{W}) \geq \frac{\text{iSNR}}{1 + \text{iSNR}}$$

and with the help of (7.10), we easily get

$$\tilde{J}(\mathbf{H}_\text{W}) \leq \frac{\text{iSNR}}{1 + \text{iSNR}}.$$

Since

$$\rho^2(\mathbf{x}, \mathbf{z}_\text{W}) \leq \frac{\text{oSNR}(\mathbf{H}_\text{W})}{1 + \text{oSNR}(\mathbf{H}_\text{W})}$$

and, again, with the help of (7.10), we obtain

$$\frac{\text{iSNR}}{1 + \text{oSNR}(\mathbf{H}_\text{W})} \leq \tilde{J}(\mathbf{H}_\text{W}).$$

Hence, we get better bounds for the MNMSE than the usual ones $[0 \leq \tilde{J}(\mathbf{H}_\text{W}) \leq 1]$.

Property 7.4. We have

$$\frac{[1 + \text{oSNR}(\mathbf{H}_\text{W})]^2}{\text{iSNR} \cdot \text{oSNR}(\mathbf{H}_\text{W})} \leq \xi_\text{nr}(\mathbf{H}_\text{W}) \leq \frac{(1 + \text{iSNR})[1 + \text{oSNR}(\mathbf{H}_\text{W})]}{\text{iSNR}^2}. \tag{7.13}$$

Proof. Easy to show by using (7.9) and the bounds of $\tilde{J}(\mathbf{H}_\text{W})$ [eq. (7.12)].

Property 7.5. We have

$$\frac{1}{[1 + \text{oSNR}(\mathbf{H}_\text{W})]^2} \leq \upsilon_\text{sd}(\mathbf{H}_\text{W}) \leq \frac{1 + \text{oSNR}(\mathbf{H}_\text{W}) - \text{iSNR}}{(1 + \text{iSNR})[1 + \text{oSNR}(\mathbf{H}_\text{W})]}. \tag{7.14}$$

Proof. Easy to show by using (7.8) and the bounds of $\xi_\text{nr}(\mathbf{H}_\text{W})$ [eq. (7.13)].

Particular case: white noise.

We assume here that the noise picked up by the microphone is white (i.e., $\mathbf{R}_v = \sigma_v^2 \mathbf{I}$). In this situation, the Wiener filter matrix becomes

$$\mathbf{H}_\text{W} = \mathbf{I} - \sigma_v^2 \mathbf{R}_y^{-1}, \tag{7.15}$$

where

$$\mathbf{R}_y = \mathbf{R}_x + \sigma_v^2 \mathbf{I}.$$

It is well known that the inverse of the Toeplitz matrix $\mathbf{R}_y$ can be factorized as follows [8], [82]:

$$\mathbf{R}_y^{-1} = \begin{bmatrix} 1 & -c_{21} & \cdots & -c_{L1} \\ -c_{12} & 1 & \cdots & -c_{L2} \\ \vdots & \vdots & \ddots & \vdots \\ -c_{1L} & -c_{2L} & \cdots & 1 \end{bmatrix} \begin{bmatrix} 1/E_1 & 0 & \cdots & 0 \\ 0 & 1/E_2 & \cdots & 0 \\ \vdots & \vdots & \ddots & \vdots \\ 0 & 0 & \cdots & 1/E_L \end{bmatrix}, \qquad (7.16)$$

where the columns of the first matrix in the right-hand side of (7.16) are the linear interpolators of the signal $y(k)$ and the elements E_l in the diagonal matrix are the respective interpolation-error powers.

Using the factorization of $\mathbf{R}_y^{-1}$ in (7.6) and (7.7), the MMSE and MNMSE can be rewritten, respectively, as

$$J(\mathbf{H}_{\mathrm{W}}) = L\sigma_v^2 - \left(\sigma_v^2\right)^2 \sum_{l=1}^{L} \frac{1}{E_l}, \qquad (7.17)$$

$$\tilde{J}(\mathbf{H}_{\mathrm{W}}) = 1 - \frac{\sigma_v^2}{L} \sum_{l=1}^{L} \frac{1}{E_l}. \qquad (7.18)$$

Assume that the noise-free speech signal, $x(k)$, is very well predictable. In this scenario, $E_l \approx \sigma_v^2$, $\forall\, l$, and replacing this value in (7.18) we find that $\tilde{J}(\mathbf{H}_{\mathrm{W}}) \approx 0$. From (4.8), we then deduce that $v_{\mathrm{sd}}(\mathbf{H}_{\mathrm{W}}) \approx 0$ (no speech distortion) and $\xi_{\mathrm{nr}}(\mathbf{H}_{\mathrm{W}}) \approx \infty$ (infinite noise reduction). Notice that, from a theoretical point of view (and with white noise), this result is independent of the SNR. Also

$$\mathbf{H}_{\mathrm{W}} \approx \begin{bmatrix} 0 & c_{12} & \cdots & c_{1L} \\ c_{21} & 0 & \cdots & c_{2L} \\ \vdots & \vdots & \ddots & \vdots \\ c_{L1} & c_{L2} & \cdots & 0 \end{bmatrix} \qquad (7.19)$$

and since $\mathbf{H}_{\mathrm{W}}\mathbf{x}(k) \approx \mathbf{x}(k)$, this means that $\xi_{\mathrm{sr}}(\mathbf{H}_{\mathrm{W}}) \approx 1$; as a result oSNR$(\mathbf{H}_{\mathrm{W}}) \approx \infty$ and we can almost perfectly recover the signal $x(k)$.

At the other extreme case, let us see now what happens when the signal of interest $x(k)$ is not predictable at all. In this situation, $E_l \approx \sigma_y^2$, $\forall\, l$ and $c_{ij} \approx 0$, $\forall\, i, j$, $i \neq j$. Using these values, we get

$$\mathbf{H}_{\mathrm{W}} \approx \frac{\mathrm{iSNR}}{1 + \mathrm{iSNR}}\mathbf{I}, \qquad (7.20)$$

$$\tilde{J}(\mathbf{H}_{\mathrm{W}}) \approx \frac{\mathrm{iSNR}}{1 + \mathrm{iSNR}}. \qquad (7.21)$$

With the help of the two previous equations, it's straightforward to obtain

64   7 Optimal Filters in the Time Domain

$$\xi_{nr}(\mathbf{H_W}) \approx \left(1 + \frac{1}{iSNR}\right)^2, \tag{7.22}$$

$$\upsilon_{sd}(\mathbf{H_W}) \approx \frac{1}{(1 + iSNR)^2}, \tag{7.23}$$

$$SNR(\mathbf{H_W}) \approx iSNR. \tag{7.24}$$

While some noise reduction is achieved (at the price of speech distortion), there is no improvement in the output SNR, meaning that the Wiener filter has no positive effect on the microphone signal $y(k)$.

This analysis, even though simple, is quite insightful. It shows that the Wiener filter may not be that bad after all, as long as the desired signal is somewhat predictable. However, in practice some discontinuities could be heard from a voiced signal to an unvoiced one, since for the former the noise will be mostly removed while it will not for the latter.

7.2 Tradeoff Filters

The time-domain NMSE as shown in (4.8) is the sum of two terms. One depends on the speech distortion while the other one depends on the noise reduction. Instead of minimizing the NMSE with respect to $\mathbf{H}$ as we already did to find the Wiener filter, we can minimize the speech-distortion index with the constraint that the noise-reduction factor is equal to a value that is greater than one. Mathematically, this is equivalent to

$$\min_{\mathbf{H}} J_x(\mathbf{H}) \quad \text{subject to} \quad J_v(\mathbf{H}) = \beta \cdot \text{tr}\,(\mathbf{R}_v), \tag{7.25}$$

where $0 < \beta < 1$ in order to have some noise reduction. If we use a Lagrange multiplier, μ, to adjoin the constraint to the cost function, (7.25) can be rewritten as

$$\mathbf{H_T} = \arg\min_{\mathbf{H}} \mathcal{L}(\mathbf{H}, \mu), \tag{7.26}$$

with

$$\mathcal{L}(\mathbf{H}, \mu) = J_x(\mathbf{H}) + \mu\,[J_v(\mathbf{H}) - \beta \cdot \text{tr}\,(\mathbf{R}_v)] \tag{7.27}$$

and $\mu \geq 0$. From (7.26), we can easily derive the optimal filter:

$$\begin{aligned}
\mathbf{H_T^T} &= (\mathbf{R}_x + \mu\mathbf{R}_v)^{-1}\,\mathbf{R}_x \\
&= [\mathbf{R}_y + (\mu - 1)\mathbf{R}_v]^{-1}\,[\mathbf{R}_y - \mathbf{R}_v] \\
&= \left[(1 - \mu)\mathbf{I} + \mu\left(\mathbf{H_W^T}\right)^{-1}\right]^{-1},
\end{aligned} \tag{7.28}$$

where the Lagrange multiplier, μ, satisfies $J_v(\mathbf{H_T}) = \beta \cdot \text{tr}\,(\mathbf{R}_v)$, which implies that

$$\xi_{\mathrm{nr}}(\mathbf{H}_{\mathrm{T}}) = \frac{1}{\beta} > 1. \tag{7.29}$$

From (4.8), we get

$$\upsilon_{\mathrm{sd}}(\mathbf{H}_{\mathrm{T}}) = \frac{1}{\mathrm{iSNR}} \left[\tilde{J}(\mathbf{H}_{\mathrm{T}}) - \beta \right], \tag{7.30}$$

and because $\upsilon_{\mathrm{sd}}(\mathbf{H}_{\mathrm{T}}) \geq 0$, $\tilde{J}(\mathbf{H}_{\mathrm{T}}) \geq \beta$.

Since $\tilde{J}(\mathbf{H}_{\mathrm{W}}) \leq \tilde{J}(\mathbf{H}_{\mathrm{T}})$, $\forall \mu$, we also have

$$\upsilon_{\mathrm{sd}}(\mathbf{H}_{\mathrm{T}}) \geq \upsilon_{\mathrm{sd}}(\mathbf{H}_{\mathrm{W}}) + \frac{1}{\mathrm{iSNR}} \left[\frac{1}{\xi_{\mathrm{nr}}(\mathbf{H}_{\mathrm{W}})} - \frac{1}{\xi_{\mathrm{nr}}(\mathbf{H}_{\mathrm{T}})} \right]. \tag{7.31}$$

Therefore, $\xi_{\mathrm{nr}}(\mathbf{H}_{\mathrm{T}}) \geq \xi_{\mathrm{nr}}(\mathbf{H}_{\mathrm{W}})$ implies that $\upsilon_{\mathrm{sd}}(\mathbf{H}_{\mathrm{T}}) \geq \upsilon_{\mathrm{sd}}(\mathbf{H}_{\mathrm{W}})$.

In practice it's not easy to determine the optimal μ. Therefore, when this parameter is chosen in an ad-hoc way, we can see that for

- $\mu = 1$, $\mathbf{H}_{\mathrm{T}} = \mathbf{H}_{\mathrm{W}}$: Wiener filter;
- $\mu = 0$, $\mathbf{H}_{\mathrm{T}} = \mathbf{I}$: distortionless filter;
- $\mu > 1$, results in low residual noise at the expense of high speech distortion;
- $\mu < 1$, we get little speech distortion but not so much noise reduction.

Property 7.6. With the tradeoff filter given in (7.28), the output SNR is always greater than or equal to the input SNR, i.e., oSNR$(\mathbf{H}_{\mathrm{T}}) \geq$ iSNR.

Proof. The SPCC between the two vectors $\mathbf{x}(k)$ and $\mathbf{x}(k) + \sqrt{\mu}\mathbf{v}(k)$ is

$$\begin{aligned} \rho^2 \left(\mathbf{x}, \mathbf{x} + \sqrt{\mu}\mathbf{v} \right) &= \frac{\mathrm{tr}^2 \left(\mathbf{R}_x \right)}{\mathrm{tr} \left(\mathbf{R}_x \right) \left[\mathrm{tr} \left(\mathbf{R}_x \right) + \mu \mathrm{tr} \left(\mathbf{R}_v \right) \right]} \\ &= \frac{\mathrm{iSNR}}{\mu + \mathrm{iSNR}}. \end{aligned}$$

The SPCC between the two vectors $\mathbf{x}(k)$ and $\mathbf{H}_{\mathrm{T}}\mathbf{x}(k) + \sqrt{\mu}\mathbf{H}_{\mathrm{T}}\mathbf{v}(k)$ is

$$\begin{aligned} \rho^2 \left(\mathbf{x}, \mathbf{H}_{\mathrm{T}}\mathbf{x} + \sqrt{\mu}\mathbf{H}_{\mathrm{T}}\mathbf{v} \right) &= \frac{\mathrm{tr}^2 \left(\mathbf{H}_{\mathrm{T}}\mathbf{R}_x \right)}{\mathrm{tr} \left(\mathbf{R}_x \right) \mathrm{tr} \left[\mathbf{H}_{\mathrm{T}} \left(\mathbf{R}_x + \mu \mathbf{R}_v \right) \mathbf{H}_{\mathrm{T}}^T \right]} \\ &= \frac{\mathrm{tr} \left(\mathbf{H}_{\mathrm{T}}\mathbf{R}_x \right)}{\mathrm{tr} \left(\mathbf{R}_x \right)}. \end{aligned}$$

Another way to write the same SPCC is the following:

$$\begin{aligned} \rho^2 \left(\mathbf{x}, \mathbf{H}_{\mathrm{T}}\mathbf{x} + \sqrt{\mu}\mathbf{H}_{\mathrm{T}}\mathbf{v} \right) &= \frac{\mathrm{tr}^2 \left(\mathbf{H}_{\mathrm{T}}\mathbf{R}_x \right)}{\mathrm{tr} \left(\mathbf{R}_x \right) \mathrm{tr} \left(\mathbf{H}_{\mathrm{T}}\mathbf{R}_x\mathbf{H}_{\mathrm{T}}^T \right)} \cdot \frac{\mathrm{oSNR} \left(\mathbf{H}_{\mathrm{T}} \right)}{\mu + \mathrm{oSNR} \left(\mathbf{H}_{\mathrm{T}} \right)} \\ &= \rho^2 \left(\mathbf{x}, \mathbf{H}_{\mathrm{T}}\mathbf{x} \right) \cdot \rho^2 \left(\mathbf{H}_{\mathrm{T}}\mathbf{x}, \mathbf{H}_{\mathrm{T}}\mathbf{x} + \sqrt{\mu}\mathbf{H}_{\mathrm{T}}\mathbf{v} \right) \\ &\leq \frac{\mathrm{oSNR} \left(\mathbf{H}_{\mathrm{T}} \right)}{\mu + \mathrm{oSNR} \left(\mathbf{H}_{\mathrm{T}} \right)}. \end{aligned}$$

66   7 Optimal Filters in the Time Domain

Now, let us evaluate the SPCC between the two vectors $\mathbf{x}(k) + \sqrt{\mu}\mathbf{v}(k)$ and $\mathbf{H}_T\mathbf{x}(k) + \sqrt{\mu}\mathbf{H}_T\mathbf{v}(k)$:

$$\rho^2\left(\mathbf{x} + \sqrt{\mu}\mathbf{v}, \mathbf{H}_T\mathbf{x} + \sqrt{\mu}\mathbf{H}_T\mathbf{v}\right) =$$
$$\frac{\mathrm{tr}^2\left[\mathbf{H}_T\left(\mathbf{R}_x + \mu\mathbf{R}_v\right)\right]}{\left[\mathrm{tr}\left(\mathbf{R}_x\right) + \mu\mathrm{tr}\left(\mathbf{R}_v\right)\right] \cdot \mathrm{tr}\left[\mathbf{H}_T\left(\mathbf{R}_x + \mu\mathbf{R}_v\right)\mathbf{H}_T^T\right]}$$
$$= \frac{\mathrm{tr}\left(\mathbf{R}_x\right)}{\mathrm{tr}\left(\mathbf{R}_x\right) + \mu\mathrm{tr}\left(\mathbf{R}_v\right)} \cdot \frac{\mathrm{tr}\left(\mathbf{R}_x\right)}{\mathrm{tr}\left(\mathbf{H}_T\mathbf{R}_x\right)}$$
$$= \frac{\rho^2\left(\mathbf{x}, \mathbf{x} + \sqrt{\mu}\mathbf{v}\right)}{\rho^2\left(\mathbf{x}, \mathbf{H}_T\mathbf{x} + \sqrt{\mu}\mathbf{H}_T\mathbf{v}\right)}.$$

Therefore

$$\rho^2\left(\mathbf{x}, \mathbf{x} + \sqrt{\mu}\mathbf{v}\right) = \frac{\mathrm{iSNR}}{\mu + \mathrm{iSNR}}$$
$$= \rho^2\left(\mathbf{x} + \sqrt{\mu}\mathbf{v}, \mathbf{H}_T\mathbf{x} + \sqrt{\mu}\mathbf{H}_T\mathbf{v}\right) \cdot \rho^2\left(\mathbf{x}, \mathbf{H}_T\mathbf{x} + \sqrt{\mu}\mathbf{H}_T\mathbf{v}\right)$$
$$\leq \rho^2\left(\mathbf{x}, \mathbf{H}_T\mathbf{x} + \sqrt{\mu}\mathbf{H}_T\mathbf{v}\right)$$
$$\leq \frac{\mathrm{oSNR}\left(\mathbf{H}_T\right)}{\mu + \mathrm{oSNR}\left(\mathbf{H}_T\right)}.$$

As a result

$$\mathrm{oSNR}\left(\mathbf{H}_T\right) \geq \mathrm{iSNR}.$$

We can find another tradeoff filter by minimizing the residual noise with the constraint that we allow some level of speech distortion. Mathematically, this is equivalent to

$$\min_{\mathbf{H}} J_v(\mathbf{H}) \quad \text{subject to} \quad J_x(\mathbf{H}) = \beta_2 \cdot \mathrm{tr}\left(\mathbf{R}_x\right), \tag{7.32}$$

where $\beta_2 > 0$ in order to have some noise reduction. If we use a Lagrange multiplier, μ_2, to adjoin the constraint to the cost function, (7.32) can be rewritten as

$$\mathbf{H}_{T,2} = \arg\min_{\mathbf{H}} \mathcal{L}(\mathbf{H}, \mu_2), \tag{7.33}$$

with

$$\mathcal{L}(\mathbf{H}, \mu_2) = J_v\left(\mathbf{H}\right) + \mu_2\left[J_x\left(\mathbf{H}\right) - \beta_2 \cdot \mathrm{tr}\left(\mathbf{R}_x\right)\right] \tag{7.34}$$

and $\mu_2 \geq 0$. The optimal solution to this optimization problem is

$$\mathbf{H}_{T,2}^T = \left(\mathbf{R}_x + \frac{\mathbf{R}_v}{\mu_2}\right)^{-1}\mathbf{R}_x, \tag{7.35}$$

where the Lagrange multiplier, μ_2, satisfies $J_x\left(\mathbf{H}_{\mathrm{T},2}\right) = \beta_2 \cdot \mathrm{tr}\left(\mathbf{R}_x\right)$, which implies that

$$\upsilon_{\mathrm{sd}}(\mathbf{H}_{\mathrm{T},2}) = \beta_2 > 0. \tag{7.36}$$

From (4.8), we get

$$\xi_{\mathrm{nr}}\left(\mathbf{H}_{\mathrm{T},2}\right) = \frac{1}{\tilde{J}(\mathbf{H}_{\mathrm{T},2}) - \beta_2 \cdot \mathrm{iSNR}}, \tag{7.37}$$

and because $\xi_{\mathrm{nr}}\left(\mathbf{H}_{\mathrm{T},2}\right) > 0$, $\tilde{J}(\mathbf{H}_{\mathrm{T},2}) > \beta_2 \cdot \mathrm{iSNR}$.

Since $\tilde{J}(\mathbf{H}_{\mathrm{W}}) \leq \tilde{J}(\mathbf{H}_{\mathrm{T},2})$, $\forall \mu$, we also have

$$\frac{1}{\xi_{\mathrm{nr}}(\mathbf{H}_{\mathrm{T},2})} \geq \frac{1}{\xi_{\mathrm{nr}}(\mathbf{H}_{\mathrm{W}})} + \mathrm{iSNR}\left[\upsilon_{\mathrm{sd}}(\mathbf{H}_{\mathrm{W}}) - \upsilon_{\mathrm{sd}}(\mathbf{H}_{\mathrm{T},2})\right]. \tag{7.38}$$

Therefore, $\upsilon_{\mathrm{sd}}(\mathbf{H}_{\mathrm{T},2}) \leq \upsilon_{\mathrm{sd}}(\mathbf{H}_{\mathrm{W}})$ implies that $\xi_{\mathrm{nr}}(\mathbf{H}_{\mathrm{T},2}) \leq \xi_{\mathrm{nr}}(\mathbf{H}_{\mathrm{W}})$.

From a practical point of view, the two tradeoff filters derived here are fundamentally the same since by taking $\mu = 1/\mu_2$, we see that $\mathbf{H}_{\mathrm{T}} = \mathbf{H}_{\mathrm{T},2}$.

7.3 Subspace Approach

In [54], it is shown that two symmetric matrices $\mathbf{R}_x$ and $\mathbf{R}_v$ can be jointly diagonalized if $\mathbf{R}_v$ is positive definite. This joint diagonalization was first introduced by Jensen et al. [78] and then by Hu and Loizou [66], [67], [68] in the single-channel noise reduction problem. For our time-domain model we get

$$\mathbf{R}_x = \mathbf{B}^T \mathbf{\Lambda}_{\mathrm{jd}} \mathbf{B}, \tag{7.39a}$$

$$\mathbf{R}_v = \mathbf{B}^T \mathbf{B}, \tag{7.39b}$$

$$\mathbf{R}_y = \mathbf{B}^T \left[\mathbf{I} + \mathbf{\Lambda}_{\mathrm{jd}}\right] \mathbf{B}, \tag{7.39c}$$

where $\mathbf{B}$ is a full rank square matrix but not necessarily orthogonal, and the diagonal matrix

$$\mathbf{\Lambda}_{\mathrm{jd}} = \mathrm{diag}\left[\lambda_{\mathrm{jd},1}\ \lambda_{\mathrm{jd},2}\ \cdots\ \lambda_{\mathrm{jd},L}\right] \tag{7.40}$$

are the eigenvalues of the matrix $\mathbf{R}_v^{-1}\mathbf{R}_x$ with $\lambda_{\mathrm{jd},1} \geq \lambda_{\mathrm{jd},2} \geq \cdots \geq \lambda_{\mathrm{jd},L} \geq 0$.

Applying the decompositions (7.39a)–(7.39c) in (7.28), the tradeoff filter becomes

$$\mathbf{H}_{\mathrm{T}} = \mathbf{B}^T \mathbf{\Lambda}_{\mathrm{jd}} \left(\mathbf{\Lambda}_{\mathrm{jd}} + \mu \mathbf{I}\right)^{-1} \mathbf{B}^{-T}. \tag{7.41}$$

Therefore, the estimation of the speech signal, $\mathbf{x}(k)$, is done in three steps: first we apply the transform $\mathbf{B}^{-T}$ to the noisy signal; second the transformed signal

68   7 Optimal Filters in the Time Domain

is modified by the gain function $\mathbf{\Lambda}_{\mathrm{jd}} \left(\mathbf{\Lambda}_{\mathrm{jd}} + \mu \mathbf{I} \right)^{-1}$; and finally we transform back the signal to its original domain by applying the transform $\mathbf{B}^T$.

Usually, a speech signal can be modelled as a linear combination of a number of some (linearly independent) basis vectors smaller than the dimension of these vectors [49], [64], [76]. As a result, the vector space of the noisy signal can be decomposed in two subspaces: the signal-plus-noise subspace of length L_{s} and the noise subspace of length L_{n}, with $L = L_{\mathrm{s}} + L_{\mathrm{n}}$. This implies that the last L_{n} eigenvalues of the matrix $\mathbf{R}_v^{-1} \mathbf{R}_x$ are equal to zero. Therefore, we can rewrite (7.41) to obtain the subspace filter:

$$\mathbf{H}_{\mathrm{S}} = \mathbf{B}^T \begin{bmatrix} \mathbf{\Sigma} & \mathbf{0}_{L_{\mathrm{s}} \times L_{\mathrm{n}}} \\ \mathbf{0}_{L_{\mathrm{n}} \times L_{\mathrm{s}}} & \mathbf{0}_{L_{\mathrm{n}} \times L_{\mathrm{n}}} \end{bmatrix} \mathbf{B}^{-T}, \tag{7.42}$$

where

$$\mathbf{\Sigma} = \mathrm{diag} \begin{bmatrix} \frac{\lambda_{\mathrm{jd},1}}{\lambda_{\mathrm{jd},1} + \mu} & \frac{\lambda_{\mathrm{jd},2}}{\lambda_{\mathrm{jd},2} + \mu} & \cdots & \frac{\lambda_{\mathrm{jd},L_{\mathrm{s}}}}{\lambda_{\mathrm{jd},L_{\mathrm{s}}} + \mu} \end{bmatrix} \tag{7.43}$$

is an $L_{\mathrm{s}} \times L_{\mathrm{s}}$ diagonal matrix. We now clearly see that noise reduction with the subspace method is achieved by nulling the noise subspace and cleaning the speech-plus-noise subspace via a reweighted reconstruction. Some noise reduction can be achieved by only nulling the noise subspace and leaving intact the speech-plus-noise subspace ($\mu = 0$). This algorithm was developed in [37].

7.4 Experiments

Having formulated different noise reduction filters in the time domain, we are now ready to evaluate their performance. A number of experiments have been carried out in various noise and operation conditions. In this section, we will present some results, which illustrate the impact of some important parameters on the performance and highlight the merits and limitations inherent in these noise reduction algorithms.

7.4.1 Experimental Setup

The clean speech signal used in our experiments was recorded from a female speaker in a quiet office environment. It was sampled at 8 kHz and quantized with 16 bits (2 bytes). The overall length of the signal is 30 s. The first 5 s of this signal and its spectrogram is visualized in Fig. 7.1. The noisy speech is obtained by adding noise to the clean speech where the noise signal is properly scaled to control the input SNR. We considered three types of noise: computer generated white Gaussian noise, car noise, and babbling noise.

The car noise is recorded in a car compartment with the car running at 55 miles per hour on a highway. The resulting signal is also digitized with

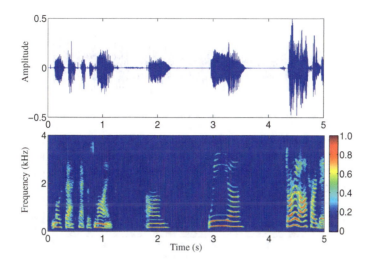

Fig. 7.1. Clean speech (first 5 s) and its spectrogram.

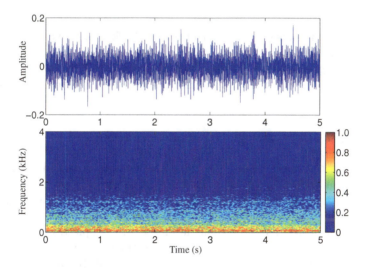

Fig. 7.2. Waveform (first 5 s) and spectrogram of car noise.

a sampling rate of 8 kHz. Figure 7.2 plots the first 5 s of this noise and its spectrogram. Compared with the white Gaussian noise which has a constant spectral energy at all frequencies, the car noise is colored in nature, and most of its energy is concentrated in low frequencies (below 2 kHz). In general, it is less stationary than the white Gaussian noise and its characteristics may change according to the road condition, running speed, and many other factors.

70   7 Optimal Filters in the Time Domain

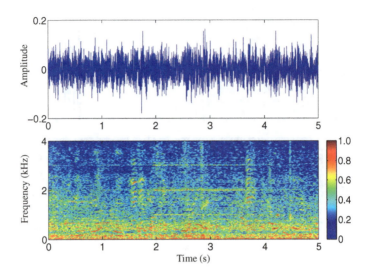

Fig. 7.3. Waveform (first 5 s) and spectrogram of NYSE babbling noise.

The babbling noise is recorded in a New York Stock Exchange (NYSE) room. This noise consists of sounds from various sources such as electric fans, telephone rings, and even speakers. As illustrated in Fig. 7.3, this noise is neither stationary nor white and its statistics change significantly from time to time.

7.4.2 Effect of Forgetting Factor on Performance

To compute the noise reduction time-domain filters, we need to know the covariance matrices $\mathbf{R}_y$ and $\mathbf{R}_v$. Since the noisy signal is accessible, the covariance matrix $\mathbf{R}_y$ can be estimated from its definition given in (2.5a) by approximating the mathematical expectation with sample average. However, due to the fact that speech is nonstationary, the sample average has to be performed on a short-term basis so that the estimated covariance matrix can follow the short-term variations of the speech signal. In speech processing, one of the most popularly used approaches to estimating $\mathbf{R}_y$ that takes into account the speech nonstationarity is the recursive method, where an estimate of $\mathbf{R}_y$ at time instant k is obtained as

$$\mathbf{R}_y(k) = \alpha_y \mathbf{R}_y(k-1) + (1-\alpha_y)\mathbf{y}(k)\mathbf{y}^T(k), \qquad (7.44)$$

where α_y is a forgetting factor that controls the influence of the previous data samples on the current estimate of the noisy signal covariance matrix. This method will be adopted in this experiment.

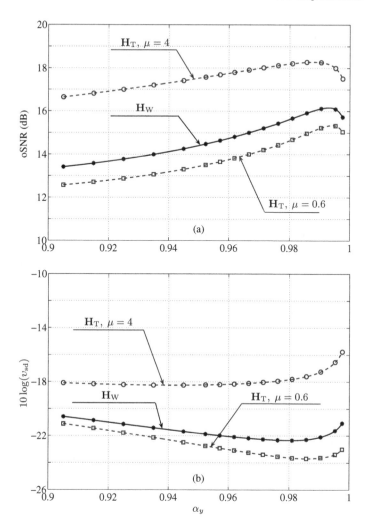

Fig. 7.4. Performance versus the forgetting factor α_y for different filters in white Gaussian noise with an iSNR = 10 dB and $L = 20$. (a) Output SNR and (b) speech-distortion index.

The computation of the noise signal covariance matrix $\mathbf{R}_v$ requires a noise estimator, which can be achieved in different ways and will be addressed in the next chapter.

This experiment aims at studying the effect of α_y on noise reduction performance. We consider the stationary white Gaussian noise and compute the matrix $\mathbf{R}_v$ directly from the noise signal using a long-term sample average.

72   7 Optimal Filters in the Time Domain

The results of this experiment[1] is plotted in Fig. 7.4 where the output SNR and speech-distortion index are computed according to (3.9) and (3.23) respectively [at each time instant, the noise reduction filter is estimated and then directly applied to the clean speech and noise to obtain $\mathbf{x}_F(k)$ and $\mathbf{v}_F(k)$, and the output SNR and speech-distortion index are subsequently computed]. It is seen from Fig. 7.4 that both the output SNR and speech-distortion index bear a nonmonotonic relationship with the forgetting factor α_y. Specifically, the output SNR first increases as α_y increases and then decreases. In comparison, the speech-distortion index first decreases with α_y and then increases. Apparently, the value of α_y plays a critical role on the noise reduction performance of the studied filters. On the one hand, the forgetting factor α_y cannot be small; otherwise, the estimation variance of $\mathbf{R}_y(k)$ will be large, which can lead to performance degradation in noise reduction. Furthermore, $\mathbf{R}_y(k)$ may tend to be rank deficient, causing numerical stability problems. On the other hand, this forgetting factor α_y cannot be too large. If it is too large (close to 1), the recursive estimate will essentially be a long-term average and will not be able to follow the short-term variations of the speech signal. As a result, the nature of the speech is not fully taken advantage of, which limits the noise reduction performance. From Fig. 7.4, we can see that the optimal noise reduction performance (highest output SNR and lowest speech distortion) occurs when α_y is between 0.985 and 0.995. So, in the subsequent experiments, we will set α_y to 0.985.

7.4.3 Effect of Filter Length on Performance

Another important parameter for all the time-domain noise reduction filters is the filter length (also the frame size) L. So, in this second experiment, we study the impact of the filter length L on the performance of noise reduction. Again, the noise is white Gaussian with an iSNR = 10 dB. Based on the previous experiment, we set $\alpha_y = 0.985$ and the noise covariance matrix $\mathbf{R}_v$ is computed using a long-term average. The results are depicted in Fig. 7.5. It is clear that the length L should be reasonably large enough to achieve good noise reduction performance. When L increases from 1 to 20, the output SNR improves while speech distortion decreases. But if we continue to increase L, there is either marginal additional SNR improvement (for the tradeoff filter with $\mu = 4$), or the output SNR even slightly degrades (for the Wiener filter and the tradeoff filter with $\mu = 0.6$), and there is also some increase in speech distortion. In general, good performance for all the studied algorithms is achieved when the filter length L is between 10 and 20. This result coincides with what was observed in [24]. The reason behind this is that speech is somewhat predictable. It is this predictability that helps us achieve noise reduction without noticeably distorting the desired speech signal. In order to

---

[1] Note that we do not present the results of the subspace method here because it does not perform better than the tradeoff filter.

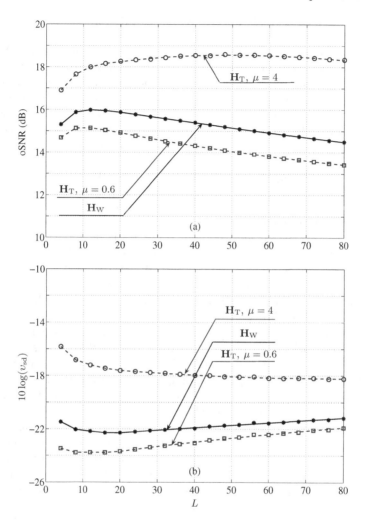

Fig. 7.5. Performance versus the filter length L for different filters in white Gaussian noise with an iSNR = 10 dB and $\alpha_y = 0.985$. (a) Output SNR and (b) speech-distortion index.

fully take advantage of the speech predictability, the filter length needs to be larger than the order of speech prediction, which is in the range between 10 and 20 for an 8-kHz sampling rate. But if we continue to increase L, the additional performance improvement will be limited. In theory, there shouldn't be performance degradation for large L. However, in practice, the estimation variance of the covariance matrix $\mathbf{R}_y$ increases with L, which generally leads to performance degradation.

74   7 Optimal Filters in the Time Domain

7.4.4 Performance in Different Noise Conditions

In this experiment, we test the performance of the different filters in various noise conditions with different input SNRs. We consider three types of noise: white Gaussian, car, and NYSE. Based on the previous experimental results, we set $L = 20$ and $\alpha_y = 0.985$. Again, we do not use any noise estimator, but compute the noise covariance directly from the noise signal. However, since we have now both stationary and nonstationary noise, we choose to compute the noise signal covariance matrix using the recursive method through

$$\mathbf{R}_v(k) = \alpha_v \mathbf{R}_v(k - 1) + (1 - \alpha_v)\mathbf{v}(k)\mathbf{v}^T(k), \qquad (7.45)$$

where $\mathbf{R}_v(k)$ is an estimate of the covariance matrix $\mathbf{R}_v$ at time k and α_v, similar to α_y, is a forgetting factor.

Theoretically, the optimal value of α_v depends on the characteristics of the noise. A large α_v should be used if the noise is stationary. But if the noise is nonstationary, the value of α_v should be set smaller. However, in this experiment, we do not assume to know the nonstationarity of each type of noise and set $\alpha_v = 0.995$ for all the three noise signals. (Note that this value of α_v is determined through experiments. We fixed α_y to 0.985 and $L = 20$, but change α_v from 0 to 1. In average, we found that the best noise reduction performance is achieved when $\alpha_v = 0.995$.) The results for this experiment are shown in Fig. 7.6, where we only plotted the results of the Wiener filter and the tradeoff filter with $\mu = 4$ to simplify the presentation.

It is seen from Fig. 7.6 that, in the same type of noise, the lower the input SNR, the more the noise is reduced (higher SNR improvement). But the speech-distortion index increases rapidly as the input SNR decreases. When the input SNR is very low (e.g., below 0 dB), the speech-distortion index becomes very large, which may cause significant negative impact to the speech signal (instead of improving the speech quality, it may degrade it due to the large speech distortion). To circumvent this problem in practical noise reduction systems, we suggest to use grace degradation, i.e., when the input SNR is above a certain threshold (around 10 dB), the noise reduction filters can be directly applied to the noisy speech; but when the input SNR is below some lower threshold (around or below 0 dB), we should leave the noisy speech unchanged; if the input SNR is between the two thresholds (we call it the grace degradation range), we can use some suboptimal filter so that there is a smooth transition in speech quality from low input SNR to high input SNR environments.

If we use the Gaussian noise as the baseline, it is seen that both the Wiener and tradeoff filters perform consistently worse in the NYSE noise. This is due to the fact that the NYSE noise is nonstationary, therefore it is more difficult to deal with. In car noise, however, both filters yielded better performance in low input SNRs but poorer performance in high input SNRs. As what caused this behavior and how to take advantage of the low-pass nature of the car

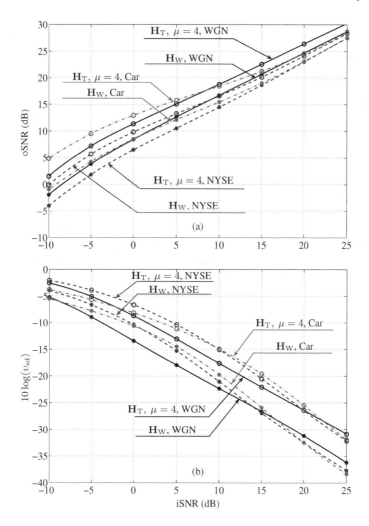

Fig. 7.6. Performance for different input SNRs and noise conditions. The parameters used in this experiment are $L = 20$, $\alpha_y = 0.985$, and $\alpha_v = 0.995$.

noise to achieve better performance will be discussed in the following two chapters.

7.5 Summary

In this chapter, we derived several forms of time-domain filters for noise reduction. The Wiener filter is optimal in the MSE sense and many known algorithms are somehow related to it. We discussed many important and interesting properties of the Wiener filter and showed that the it achieves SNR improvement at the cost of speech distortion. We also derived two tradeoff filters, which are also optimal from an MMSE point of view. The only difference between the Wiener and tradeoff filters is that the latter one is deduced from a constrained MSE criterion where a parameter is used to control the compromise between the amount of noise reduction and the amount of speech distortion. The subspace method can be viewed as a special way of implementing the tradeoff filter. Using experiments, we have also illustrated the impact of some important parameters on the performance of noise reduction time-domain filters.

While the time domain is the most straightforward and natural domain to work with, the derived noise reduction filters are often not very flexible in terms of performance tuning. For example, in practice, noise is not necessarily white and in many cases its energy may be concentrated only in some frequency bands (such as the car noise). In such situations, it is advantageous to design a noise reduction filter in the frequency domain, which will be discussed in the next chapter.

8

Optimal Filters in the Frequency Domain

We now investigate the design of noise reduction filters in the frequency do-main, which is an alternative to the time-domain approach studied in the pre-vious chapter. In the real world, frequency-domain filters are, by far, the most popular methods. This is mainly because: 1) the filters at different frequencies (or frequency bands) are designed and handled independently with each other, therefore there is a significant flexibility in dealing with colored noise; 2) most of our knowledge and understanding of speech production and perception is related to frequencies; and 3) thanks to the fast Fourier transform (FFT), the implementation of frequency-domain filters are generally very efficient. So, in this chapter, we develop some widely used classical frequency-domain filters and discuss their properties.

8.1 Wiener Filter

Taking the gradient of $J[H(j\omega)]$ [eq. (4.15)] with respect to $H^*(j\omega)$ and equating the result to 0 lead to

$$-E\left\{Y^*(j\omega)\left[X(j\omega) - H_{\mathrm{W}}(j\omega)Y(j\omega)\right]\right\} = 0. \tag{8.1}$$

Hence

$$\phi_y(\omega)H_{\mathrm{W}}(j\omega) = \phi_{xy}(j\omega). \tag{8.2}$$

But

$$\phi_{xy}(j\omega) = E\left[X(j\omega)Y^*(j\omega)\right]$$
$$= \phi_x(\omega).$$

Therefore the optimal filter can be put into the following forms:

J. Benesty et al., *Noise Reduction in Speech Processing*, Springer Topics in Signal Processing 2, DOI 10.1007/978-3-642-00296-0_8, © Springer-Verlag Berlin Heidelberg 2009

$$H_W(j\omega) = \frac{\phi_x(\omega)}{\phi_y(\omega)}$$

$$= 1 - \frac{\phi_v(\omega)}{\phi_y(\omega)}$$

$$= \frac{\mathrm{iSNR}(\omega)}{1 + \mathrm{iSNR}(\omega)}. \tag{8.3}$$

We see that the noncausal Wiener filter is always real and positive. Therefore, from now on we will drop the imaginary unit from $H_W(j\omega)$, i.e., $H_W(\omega)$, to accentuate the fact that this filter is a real number. The same frequency-domain filter can also be easily deduced from the framework of Bayesian risk theory in the particular case where the model is Gaussian [39], [58], [126].

Another way to define the Wiener filter is with the MSCFs. Indeed, it is easy to see that

$$H_W(\omega) = |\rho\,[X(j\omega), Y(j\omega)]\,|^2$$

$$= 1 - |\rho\,[V(j\omega), Y(j\omega)]\,|^2. \tag{8.4}$$

These fundamental forms of the Wiener filter, although obvious, do not seem to be known in the literature. They show that they are simply related to two MSCFs. Since $0 \le |\rho\,[X(j\omega), Y(j\omega)]\,|^2 \le 1$, therefore $0 \le H_W(\omega) \le 1$. The Wiener filter acts like a gain function. When the level of noise is high at frequency ω, $|\rho\,[V(j\omega), Y(j\omega)]\,|^2 \approx 1$, then $H_W(\omega)$ is close to 0 since there is a large amount of noise that has to be removed. When the level of noise is low at frequency ω, $|\rho\,[V(j\omega), Y(j\omega)]\,|^2 \approx 0$, then $H_W(\omega)$ is close to 1 and is not going to affect much the signals since there is little noise that needs to be removed.

Now, let us define the complex number[1]

$$\varrho\,[X(j\omega), V(j\omega)] = \rho\,[X(j\omega), Y(j\omega)] + j\rho\,[V(j\omega), Y(j\omega)]$$

$$= \cos\theta(\omega) + j\sin\theta(\omega), \tag{8.5}$$

where $\theta(\omega)$ is the angle of $\varrho\,[X(j\omega), V(j\omega)]$ for which the modulus is equal to 1. On the complex plane, $\varrho\,[X(j\omega), V(j\omega)]$ is on the unit circle. Since $0 \le \rho\,[X(j\omega), Y(j\omega)] \le 1$ and $0 \le \rho\,[V(j\omega), Y(j\omega)] \le 1$, therefore $0 \le \theta(\omega) \le \frac{\pi}{2}$. We can then rewrite the Wiener filter as a function of the angle $\theta(\omega)$:

$$H_W(\omega) = \cos^2\theta(\omega)$$

$$= 1 - \sin^2\theta(\omega). \tag{8.6}$$

Hence

$$\lim_{\theta(\omega) \to 0} H_W(\omega) = 1, \tag{8.7}$$

$$\lim_{\theta(\omega) \to \frac{\pi}{2}} H_W(\omega) = 0. \tag{8.8}$$

---

[1] Notice that both $\rho\,[X(j\omega), Y(j\omega)]$ and $\rho\,[V(j\omega), Y(j\omega)]$ are real numbers.

We deduce the subband noise-reduction factor and subband speech-distortion index

$$\xi_{\mathrm{nr}}\left[H_{\mathrm{W}}(\omega)\right] = \frac{1}{\cos^4 \theta(\omega)} \geq 1, \tag{8.9}$$

$$v_{\mathrm{sd}}\left[H_{\mathrm{W}}(\omega)\right] = \sin^4 \theta(\omega) \leq 1, \tag{8.10}$$

and the subband MNMSE

$$\tilde{J}\left[H_{\mathrm{W}}(\omega)\right] = H_{\mathrm{W}}(\omega), \tag{8.11}$$

which is exactly the Wiener filter. We see clearly how noise reduction and speech distortion depend on the angle $\theta(\omega)$ in the noncausal Wiener filter. When $\theta(\omega)$ increases so does $\xi_{\mathrm{nr}}\left[H_{\mathrm{W}}(\omega)\right]$; at the same time $v_{\mathrm{sd}}\left[H_{\mathrm{W}}(\omega)\right]$ increases.

Property 8.1. With the optimal noncausal Wiener filter given in (8.3), the fullband output SNR is always greater than or equal to the input SNR, i.e., $\mathrm{oSNR}(H_{\mathrm{W}}) \geq \mathrm{iSNR}$.

Proof. Let us evaluate the fullband SPCC between $Y(j\omega)$ and $Z_{\mathrm{W}}(j\omega) = H_{\mathrm{W}}(\omega)Y(j\omega)$:

$$\begin{aligned}
|\rho\left(Y, Z_{\mathrm{W}}\right)|^2 &= \frac{\left[\int_{-\pi}^{\pi} H_{\mathrm{W}}(\omega)\phi_y(\omega)d\omega\right]^2}{\left[\int_{-\pi}^{\pi} \phi_y(\omega)d\omega\right]\left[\int_{-\pi}^{\pi} H_{\mathrm{W}}^2(\omega)\phi_y(\omega)d\omega\right]} \\
&= \frac{\int_{-\pi}^{\pi} \phi_x(\omega)d\omega}{\int_{-\pi}^{\pi} \phi_y(\omega)d\omega} \cdot \frac{\int_{-\pi}^{\pi} \phi_x(\omega)d\omega}{\int_{-\pi}^{\pi} H_{\mathrm{W}}(\omega)\phi_x(\omega)d\omega} \\
&= \frac{|\rho\left(X, Y\right)|^2}{|\rho\left(X, Z_{\mathrm{W}}\right)|^2}.
\end{aligned}$$

Therefore

$$|\rho\left(X, Y\right)|^2 = |\rho\left(Y, Z_{\mathrm{W}}\right)|^2 \cdot |\rho\left(X, Z_{\mathrm{W}}\right)|^2 \leq |\rho\left(X, Z_{\mathrm{W}}\right)|^2.$$

Using (6.31) and Property 6.12 in the previous expression, we get

$$\frac{\mathrm{iSNR}}{1 + \mathrm{iSNR}} \leq \frac{\mathrm{oSNR}(H_{\mathrm{W}})}{1 + \mathrm{oSNR}(H_{\mathrm{W}})},$$

as a result

$$\mathrm{oSNR}(H_{\mathrm{W}}) \geq \mathrm{iSNR}.$$

Substituting (8.3) into (4.22), we find the fullband MNMSE:

$$\tilde{J}\left(H_{\mathrm{W}}\right) = 1 - \frac{\int_{-\pi}^{\pi} \phi_v^2(\omega)\phi_y^{-1}(\omega)d\omega}{\int_{-\pi}^{\pi} \phi_v(\omega)d\omega} \leq 1. \tag{8.12}$$

80   8 Optimal Filters in the Frequency Domain

We can compute the fullband speech-distortion index by substituting (8.3) into (3.26):

$$\upsilon_{sd}\left(H_{\mathrm{W}}\right) = 1 - \frac{\mathrm{oSNR}(H_{\mathrm{W}}) + 2}{\mathrm{iSNR} \cdot \xi_{\mathrm{nr}}\left(H_{\mathrm{W}}\right)} \leq 1. \tag{8.13}$$

Using (4.22) and (8.13), we get the fullband noise-reduction factor:

$$\xi_{\mathrm{nr}}\left(H_{\mathrm{W}}\right) = \frac{\mathrm{oSNR}(H_{\mathrm{W}}) + 1}{\mathrm{iSNR} - \tilde{J}\left(H_{\mathrm{W}}\right)} \geq 1. \tag{8.14}$$

Property 8.2. We have

$$\tilde{J}\left(H_{\mathrm{W}}\right) = \mathrm{iSNR}\left[1 - |\rho(X, Z_{\mathrm{W}})|^2\right]. \tag{8.15}$$

Proof. Indeed

$$\tilde{J}\left(H_{\mathrm{W}}\right) = \frac{\int_{-\pi}^{\pi} \phi_x(\omega)d\omega}{\int_{-\pi}^{\pi} \phi_v(\omega)d\omega} - \frac{\int_{-\pi}^{\pi} \phi_x^2(\omega)\phi_y^{-1}(\omega)d\omega}{\int_{-\pi}^{\pi} \phi_v(\omega)d\omega}$$

$$= \mathrm{iSNR}\left[1 - \frac{\int_{-\pi}^{\pi} \phi_x^2(\omega)\phi_y^{-1}(\omega)d\omega}{\int_{-\pi}^{\pi} \phi_x(\omega)d\omega}\right]$$

$$= \mathrm{iSNR}\left[1 - |\rho(X, Z_{\mathrm{W}})|^2\right].$$

Therefore, the fullband NMSE is minimized when the fullband SPCC between the signals $X(j\omega)$ and $Z(j\omega)$ is maximized. This fullband SPCC can be rewritten as follows:

$$|\rho(X, Z_{\mathrm{W}})|^2 = \frac{1}{\xi_{\mathrm{sr}}\left(H_{\mathrm{W}}\right)} \cdot \frac{1 + \mathrm{oSNR}(H_{\mathrm{W}})}{\mathrm{oSNR}(H_{\mathrm{W}})}. \tag{8.16}$$

We observe that the noncausal Wiener filter is compromising between speech reduction (i.e., speech distortion) and fullband output SNR improvement.

Property 8.3. We have

$$\frac{\mathrm{iSNR}}{1 + \mathrm{oSNR}(H_{\mathrm{W}})} \leq \tilde{J}\left(H_{\mathrm{W}}\right) \leq \frac{\mathrm{iSNR}}{1 + \mathrm{iSNR}}. \tag{8.17}$$

Proof. Since

$$|\rho(X, Z_{\mathrm{W}})|^2 \geq \frac{\mathrm{iSNR}}{1 + \mathrm{iSNR}}$$

and with the help of (8.15), we easily get

$$\tilde{J}\left(H_{\mathrm{W}}\right) \leq \frac{\mathrm{iSNR}}{1 + \mathrm{iSNR}}.$$

Since

$$|\rho(X, Z_{\mathrm{W}})|^2 \leq \frac{\mathrm{oSNR}(H_{\mathrm{W}})}{1 + \mathrm{oSNR}(H_{\mathrm{W}})}$$

and, again, with the help of (8.15), we obtain

$$\frac{\mathrm{iSNR}}{1 + \mathrm{oSNR}(H_{\mathrm{W}})} \leq \tilde{J}(H_{\mathrm{W}}).$$

Hence, we get better bounds for the fullband MNMSE than the usual ones $[0 \leq \tilde{J}(H_{\mathrm{W}}) \leq 1]$.

Property 8.4. We have

$$\frac{[1 + \mathrm{oSNR}(H_{\mathrm{W}})]^2}{\mathrm{iSNR} \cdot \mathrm{oSNR}(H_{\mathrm{W}})} \leq \xi_{\mathrm{nr}}(H_{\mathrm{W}}) \leq \frac{(1 + \mathrm{iSNR})[1 + \mathrm{oSNR}(H_{\mathrm{W}})]}{\mathrm{iSNR}^2}. \quad (8.18)$$

Proof. Easy to show by using (8.14) and the bounds of $\tilde{J}(H_{\mathrm{W}})$ [eq. (8.17)].

Property 8.5. We have

$$\frac{1}{[1 + \mathrm{oSNR}(H_{\mathrm{W}})]^2} \leq \upsilon_{\mathrm{sd}}(H_{\mathrm{W}}) \leq \frac{1 + \mathrm{oSNR}(H_{\mathrm{W}}) - \mathrm{iSNR}}{(1 + \mathrm{iSNR})[1 + \mathrm{oSNR}(H_{\mathrm{W}})]}. \quad (8.19)$$

Proof. Easy to show by using (8.13) and the bounds of $\xi_{\mathrm{nr}}(H_{\mathrm{W}})$ [eq. (8.18)].

Very often in practice, the ensemble averages are unknown, so it is convenient to approximate the PSDs used in the Wiener filter by sample estimates [38], [119]:

$$\begin{aligned}
\hat{H}_{\mathrm{W}}(\omega) &= 1 - \frac{V^2(\omega)}{Y^2(\omega)} \\
&= |\hat{\rho}[V(j\omega), Y(j\omega)]|^2.
\end{aligned} \quad (8.20)$$

This form of the Wiener filter is the starting point of so many spectrum-based noise reduction techniques [33], [44], [86], [88], [119].

8.2 Parametric Wiener Filter

Some applications may need aggressive noise reduction. Other applications on the contrary may require little speech distortion (so less aggressive noise reduction). An easy way to control the compromise between noise reduction and speech distortion is via the parametric Wiener filter[2] [50], [85]:

---

[2] There is nothing optimal about the parametric Wiener filter but for convenience of presentation we included it in this chapter.

82   8 Optimal Filters in the Frequency Domain

$$H_G(\omega) = \left[1 - \sin^{\beta_1} \theta(\omega)\right]^{\beta_2},\qquad(8.21)$$

where β_1 and β_2 are two positive parameters that allow the control of this compromise. For $(\beta_1, \beta_2) = (2, 1)$, we get the noncausal Wiener filter developed previously. Taking $(\beta_1, \beta_2) = (2, 1/2)$, leads to

$$H_P(\omega) = \sqrt{1 - \sin^2 \theta(\omega)}\qquad(8.22)$$
$$= \cos\theta(\omega),$$

which is the power subtraction method studied in [44], [50], [85], [92], [112]. The pair $(\beta_1, \beta_2) = (1, 1)$ gives the magnitude subtraction method [17], [18], [108], [109], [122]:

$$H_M(\omega) = 1 - \sin\theta(\omega)\qquad(8.23)$$
$$= 1 - \sqrt{1 - \cos^2\theta(\omega)}.$$

We can verify that the subband noise-reduction factors for the power subtraction and magnitude subtraction methods are

$$\xi_{nr}\left[H_P(\omega)\right] = \frac{1}{\cos^2\theta(\omega)},\qquad(8.24)$$

$$\xi_{nr}\left[H_M(\omega)\right] = \frac{1}{\left[1 - \sin\theta(\omega)\right]^2},\qquad(8.25)$$

and the corresponding subband speech-distortion indices are

$$\upsilon_{sd}\left[H_P(\omega)\right] = \left[1 - \cos\theta(\omega)\right]^2,\qquad(8.26)$$
$$\upsilon_{sd}\left[H_M(\omega)\right] = \sin^2\theta(\omega).\qquad(8.27)$$

We can also easily check that

$$\xi_{nr}\left[H_M(\omega)\right] \geq \xi_{nr}\left[H_W(\omega)\right] \geq \xi_{nr}\left[H_P(\omega)\right],\qquad(8.28)$$
$$\upsilon_{sd}\left[H_P(\omega)\right] \leq \upsilon_{sd}\left[H_W(\omega)\right] \leq \upsilon_{sd}\left[H_M(\omega)\right].\qquad(8.29)$$

The two previous inequalities are very important from a practical point of view. They show that, among the three methods, the magnitude subtraction is the most aggressive one as far as noise reduction is concerned, a very well-known fact in the literature [38], but at the same time it's the one that will likely distorts most the speech signal. The smoothest approach is the power subtraction while the Wiener filter is between the two others in terms of speech distortion and noise reduction. Several other variants of these algorithms can be found in [62], [87], [110].

8.3 Tradeoff Filter

An important filter can be designed by minimizing the speech distortion with the constraint that the residual noise level is equal to a positive number smaller than the level of the original noise. This optimization problem can be translated mathematically as

$$\min_{H(j\omega)} J_x\left[H(j\omega)\right] \quad \text{subject to} \quad J_v\left[H(j\omega)\right] = \beta \cdot \phi_v(\omega), \tag{8.30}$$

where

$$J_x\left[H(j\omega)\right] = \left|1 - H(j\omega)\right|^2 \phi_x(\omega), \tag{8.31}$$

$$J_v\left[H(j\omega)\right] = \left|H(j\omega)\right|^2 \phi_v(\omega), \tag{8.32}$$

and $0 < \beta < 1$ in order to have some noise reduction at frequency ω. If we use a Lagrange multiplier, $\mu \geq 0$, to adjoin the constraint to the cost function, we easily find the tradeoff filter:

$$\begin{aligned} H_{\mathrm{T}}(\omega) &= \frac{\phi_x(\omega)}{\phi_x(\omega) + \mu\phi_v(\omega)} \\ &= \frac{\phi_y(\omega) - \phi_v(\omega)}{\phi_y(\omega) + (\mu - 1)\phi_v(\omega)} \\ &= \frac{\mathrm{iSNR}(\omega)}{\mu + \mathrm{iSNR}(\omega)}. \end{aligned} \tag{8.33}$$

This filter can be seen as a noncausal Wiener filter with adjustable input noise level $\mu\phi_v(\omega)$.

The MSCF between the two signals $X(j\omega)$ and $X(j\omega) + \sqrt{\mu}V(j\omega)$ at frequency ω is

$$\left|\rho\left[X(j\omega), X(j\omega) + \sqrt{\mu}V(j\omega)\right]\right|^2 = \frac{\mathrm{iSNR}(\omega)}{\mu + \mathrm{iSNR}(\omega)}. \tag{8.34}$$

The MSCF between the two signals $V(j\omega)$ and $X(j\omega) + \sqrt{\mu}V(j\omega)$ at frequency ω is

$$\left|\rho\left[V(j\omega), X(j\omega) + \sqrt{\mu}V(j\omega)\right]\right|^2 = \frac{\mu}{\mu + \mathrm{iSNR}(\omega)}. \tag{8.35}$$

Therefore, we can write the tradeoff filter as a function of these two MSCFs:

$$\begin{aligned} H_{\mathrm{T}}(\omega) &= \left|\rho\left[X(j\omega), X(j\omega) + \sqrt{\mu}V(j\omega)\right]\right|^2 \\ &= 1 - \left|\rho\left[V(j\omega), X(j\omega) + \sqrt{\mu}V(j\omega)\right]\right|^2. \end{aligned} \tag{8.36}$$

Now, let us define the complex number[3]

$$\begin{aligned} \varrho_\mu\left[X(j\omega), V(j\omega)\right] &= \rho\left[X(j\omega), X(j\omega) + \sqrt{\mu}V(j\omega)\right] + \\ &\quad j\rho\left[V(j\omega), X(j\omega) + \sqrt{\mu}V(j\omega)\right] \\ &= \cos\theta(\omega, \mu) + j\sin\theta(\omega, \mu), \end{aligned} \tag{8.37}$$

---

[3] Notice that both $\rho\left[X(j\omega), X(j\omega) + \sqrt{\mu}V(j\omega)\right]$ and $\rho\left[V(j\omega), X(j\omega) + \sqrt{\mu}V(j\omega)\right]$ are real numbers.

84　8 Optimal Filters in the Frequency Domain

where $\theta(\omega, \mu)$ is the angle of $\varrho_\mu \left[X(j\omega), V(j\omega) \right]$ for which the modulus is equal to 1. On the complex plane, $\varrho_\mu \left[X(j\omega), V(j\omega) \right]$ is on the unit circle. Since $0 \leq \rho \left[X(j\omega), X(j\omega) + \sqrt{\mu}V(j\omega) \right] \leq 1$ and $0 \leq \rho \left[V(j\omega), X(j\omega) + \sqrt{\mu}V(j\omega) \right] \leq 1$, therefore $0 \leq \theta(\omega, \mu) \leq \frac{\pi}{2}$. We can then rewrite the tradeoff filter as a function of the angle $\theta(\omega, \mu)$:

$$\begin{aligned} H_T(\omega) &= \cos^2 \theta(\omega, \mu) \\ &= 1 - \sin^2 \theta(\omega, \mu). \end{aligned} \tag{8.38}$$

We deduce the subband noise-reduction factor and subband speech-distortion index

$$\xi_{nr} \left[H_T(\omega) \right] = \frac{1}{\cos^4 \theta(\omega, \mu)} \geq 1, \tag{8.39}$$

$$\upsilon_{sd} \left[H_T(\omega) \right] = \sin^4 \theta(\omega, \mu) \leq 1. \tag{8.40}$$

Property 8.6. With the tradeoff filter given in (8.33), the fullband output SNR is always greater than or equal to the input SNR, i.e., $\text{oSNR}(H_T) \geq \text{iSNR}$.

Proof. The fullband SPCC between the two variables $X(j\omega)$ and $X(j\omega) + \sqrt{\mu}V(j\omega)$ is

$$\begin{aligned} \left| \rho \left(X, X + \sqrt{\mu}V \right) \right|^2 &= \frac{\left[\int_{-\pi}^{\pi} \phi_x(\omega)d\omega \right]^2}{\left[\int_{-\pi}^{\pi} \phi_x(\omega)d\omega \right] \left[\int_{-\pi}^{\pi} \phi_x(\omega)d\omega + \mu \int_{-\pi}^{\pi} \phi_v(\omega)d\omega \right]} \\ &= \frac{\text{iSNR}}{\mu + \text{iSNR}}. \end{aligned}$$

The fullband SPCC between the two variables $X(j\omega)$ and $H_T(\omega)X(j\omega) + \sqrt{\mu}H_T(\omega)V(j\omega)$ is

$$\begin{aligned} \left| \rho \left(X, H_T X + \sqrt{\mu}H_T V \right) \right|^2 &= \\ &\frac{\left[\int_{-\pi}^{\pi} H_T(\omega)\phi_x(\omega)d\omega \right]^2}{\left[\int_{-\pi}^{\pi} \phi_x(\omega)d\omega \right] \left[\int_{-\pi}^{\pi} H_T^2(\omega)\phi_x(\omega)d\omega + \mu \int_{-\pi}^{\pi} H_T^2(\omega)\phi_v(\omega)d\omega \right]} \\ &= \frac{\int_{-\pi}^{\pi} H_T(\omega)\phi_x(\omega)d\omega}{\int_{-\pi}^{\pi} \phi_x(\omega)d\omega}. \end{aligned}$$

Another way to write the same fullband SPCC is the following:

$$|\rho\left(X, H_\mathrm{T}X + \sqrt{\mu}H_\mathrm{T}V\right)|^2 = \frac{\left[\int_{-\pi}^{\pi} H_\mathrm{T}(\omega)\phi_x(\omega)d\omega\right]^2}{\left[\int_{-\pi}^{\pi} \phi_x(\omega)d\omega\right]\left[\int_{-\pi}^{\pi} H_\mathrm{T}^2(\omega)\phi_x(\omega)d\omega\right]} \times$$

$$\frac{\mathrm{oSNR}\left(H_\mathrm{T}\right)}{\mu + \mathrm{oSNR}\left(H_\mathrm{T}\right)}$$

$$= |\rho\left(X, H_\mathrm{T}X\right)|^2 \cdot |\rho\left(H_\mathrm{T}X, H_\mathrm{T}X + \sqrt{\mu}H_\mathrm{T}V\right)|^2$$

$$\leq \frac{\mathrm{oSNR}\left(H_\mathrm{T}\right)}{\mu + \mathrm{oSNR}\left(H_\mathrm{T}\right)}.$$

Now, let us evaluate the fullband SPCC between the two variables $X(j\omega) + \sqrt{\mu}V(j\omega)$ and $H_\mathrm{T}(\omega)X(j\omega) + \sqrt{\mu}H_\mathrm{T}(\omega)V(j\omega)$:

$$|\rho\left(X + \sqrt{\mu}V, H_\mathrm{T}X + \sqrt{\mu}H_\mathrm{T}V\right)|^2 = \frac{\int_{-\pi}^{\pi} \phi_x(\omega)d\omega}{\int_{-\pi}^{\pi} \phi_x(\omega)d\omega + \mu \int_{-\pi}^{\pi} \phi_v(\omega)d\omega} \times$$

$$\frac{\int_{-\pi}^{\pi} \phi_x(\omega)d\omega}{\int_{-\pi}^{\pi} H_\mathrm{T}(\omega)\phi_x(\omega)d\omega}$$

$$= \frac{|\rho\left(X, X + \sqrt{\mu}V\right)|^2}{|\rho\left(X, H_\mathrm{T}X + \sqrt{\mu}H_\mathrm{T}V\right)|^2}.$$

Therefore

$$|\rho\left(X, X + \sqrt{\mu}V\right)|^2 = \frac{\mathrm{iSNR}}{\mu + \mathrm{iSNR}}$$

$$= |\rho\left(X + \sqrt{\mu}V, H_\mathrm{T}X + \sqrt{\mu}H_\mathrm{T}V\right)|^2 \times$$

$$|\rho\left(X, H_\mathrm{T}X + \sqrt{\mu}H_\mathrm{T}V\right)|^2$$

$$\leq |\rho\left(X, H_\mathrm{T}X + \sqrt{\mu}H_\mathrm{T}V\right)|^2$$

$$\leq \frac{\mathrm{oSNR}\left(H_\mathrm{T}\right)}{\mu + \mathrm{oSNR}\left(H_\mathrm{T}\right)}.$$

As a result

$$\mathrm{oSNR}\left(H_\mathrm{T}\right) \geq \mathrm{iSNR}.$$

The tradeoff filter can be more general if we make the factor β dependent on the frequency, i.e., $\beta(\omega)$. By doing so, the control between noise reduction and speech distortion can be more effective since each frequency ω can be controlled independently of the others. With this consideration, we can easily see that the optimal filter derived from the criterion (8.30) is now

$$H_\mathrm{T}(\omega) = \frac{\mathrm{iSNR}(\omega)}{\mu(\omega) + \mathrm{iSNR}(\omega)}, \tag{8.41}$$

where $\mu(\omega)$ is the frequency-dependent Lagrange multiplier. This approach can now provide some noise spectral shaping for masking by the speech signal [49], [68], [69], [75], [120], [121].

86   8 Optimal Filters in the Frequency Domain

8.4 Experiments

We have discussed in the previous sections several noise reduction filters in the frequency domain. While some filters (such as the Wiener filter) are optimal in the MMSE sense, others are constructed empirically (such as the parametric Wiener filter) and have no optimality properties associated with them, but they make good sense for managing the compromise between noise reduction and speech distortion. In this section, we study the performance of these main frequency-domain filters through experiments. The experimental setup is the same as used in the previous chapter.

8.4.1 Impact of Input SNR on Filter Gain and Speech Distortion

As seen from the previous sections, all the main frequency-domain filters, including the Wiener, parametric Wiener, and tradeoff filters, are essentially a function of the subband input SNR, i.e., iSNR(ω). Figure 8.1 plots the theoretical gain of each filter versus the input SNR at frequency ω where the gain is shown in both linear and dB scales. With no exception, we see that the gain of each filter is always in the range between 0 and 1, and its value is proportional to the subband input SNR. Since a lower gain corresponds to more noise attenuation, we can expect significant amount of noise suppression when iSNR(ω) is small while less noise reduction when iSNR(ω) is large. This is, of course, reasonable. When iSNR(ω) is small, the signal is very noisy, so there will be much noise to be suppressed. But if iSNR(ω) is large, the signal is relatively clean, so, there won't be much noise to be cleaned.

Comparatively, the speech-distortion index decreases exponentially with the increase of the input SNR as shown in Fig. 8.2. Since a higher speech-distortion index indicates more speech distortion, we can expect significant amount of speech distortion when the subband input SNR is low. Putting Figs. 8.1 and 8.2 together, one can clearly see that noise reduction is always associated with speech distortion. The more the noise is reduced, the more the speech is distorted. As for how these filters behave in practical environments, this will be studied through experiments. But first let us discuss how to estimate the noise signal so that we can implement the frequency-domain filters.

8.4.2 Noise Estimation

An important issue, in the implementation of an algorithm for noise reduction, is the estimation of the statistics of the noise signal. The accuracy of this estimator greatly affects the noise reduction performance. There has been a tremendous effort in tackling this problem. Representative approaches include estimating noise in the absence of speech, minimum statistics method, quantile-based method, and sequential estimation using single-pole recursion.

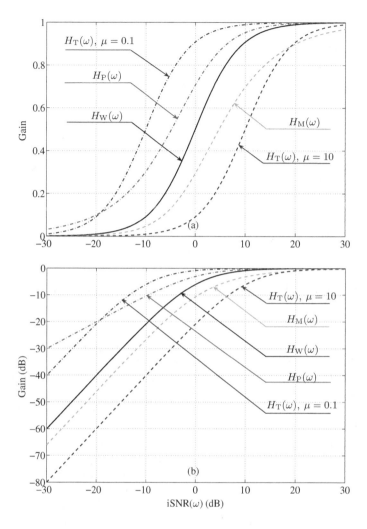

Fig. 8.1. Theoretical gain of different filters as a function of the subband input SNR, i.e., iSNR(ω). (a) Gain in linear scale and (b) Gain in dB.

Usually, a noisy speech signal is not occupied by speech all the time. In a large percentage of the time it is occupied by noise only. Therefore, a voice activity detector (VAD) can be designed to distinguish speech and non-speech segments for a given noisy signal and the noise can then be estimated from regions where the speech signal is absent. This basic noise estimation relies on a VAD with high detection accuracy. When noise is strong so that the input SNR becomes rather low, the distinction between speech and noise segments could be difficult. Moreover, noise is estimated intermittently and obtained

8 Optimal Filters in the Frequency Domain

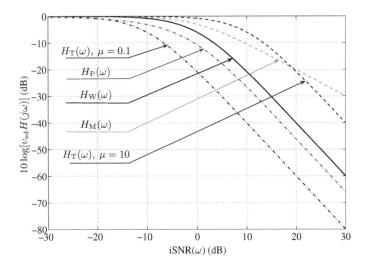

Fig. 8.2. Theoretical speech-distortion index as a function of the subband input SNR, i.e., iSNR(ω).

only during speech silent periods. This may cause problems if the noise is non-stationary, which is the case in many applications.

To avoid explicit speech/non-speech detection, the so-called minimum statistics method was developed [89]. This technique is based on the assumption that during a speech pause, or within brief periods between words and even syllables, the speech energy is close to zero. As a result, a short-term power spectrum estimate of the noisy signal, even during speech activity, decays frequently due to the noise power. Thus, by tracking the temporal spectral minimum without distinguishing between speech presence and speech absence, the noise power in a specific frequency band can be estimated. Although a VAD is not necessary in this approach, the noise estimate is often too small to provide sufficient noise reduction due to the significant variability inherent to the spectral estimation.

Instead of using minimum statistics, Hirsch et al. proposed a histogram based method which achieves a noise estimate from the subband energy histograms [65]. A threshold is set over which peaks in the histogram profile are attributed to speech. The highest peak in the profile below this threshold is treated as noise energy. This idea was extended in [114] to a quantile-based noise estimation approach, which works on the assumption that even in active speech sections of the input signal not all frequency bands are permanently occupied with speech, and for a large percentage of the time the energy is at the noise level. Thus, this method computes short-term power spectra and sorts them. The noise estimate is obtained by taking a value near to the median of the resulting profiles.

8.4 Experiments 89

Recently, sequential estimation methods [23], [38] were developed, where noise is estimated sequentially using a single-pole recursive average with an implicit speech/non-speech decision embedded. Briefly, the noisy signal $y(k)$ is segmented into blocks of L samples. Each block is then transformed via a DFT (discrete Fourier transform) into a block of L spectral samples. Successive blocks of spectral samples form a two-dimensional time-frequency matrix denoted by $Y_n(j\omega)$, where the subscript n is the frame index and denotes the time dimension. Then an estimate of the magnitude of the noise spectrum is formulated as

$$|\hat{V}_n(j\omega)| = \begin{cases} \alpha_a|\hat{V}_{n-1}(j\omega)| + (1-\alpha_a)|Y_n(j\omega)|, & \text{if } |Y_n(j\omega)| \geq |\hat{V}_{n-1}(j\omega)| \\ \alpha_d|\hat{V}_{n-1}(j\omega)| + (1-\alpha_d)|Y_n(j\omega)|, & \text{if } |Y_n(j\omega)| < |\hat{V}_{n-1}(j\omega)| \end{cases},$$
(8.42)

where α_a is the "attack" coefficient and α_d is the "decay" coefficient. Meanwhile, to reduce its temporal fluctuation, the magnitude of the noisy speech spectrum is smoothed according to the following recursion:

$$|\bar{Y}_n(j\omega)| = \begin{cases} \beta_a|\bar{Y}_{n-1}(j\omega)| + (1-\beta_a)|Y_n(j\omega)|, & \text{if } |Y_n(j\omega)| \geq |\bar{Y}_{n-1}(j\omega)| \\ \beta_d|\bar{Y}_{n-1}(j\omega)| + (1-\beta_d)|Y_n(j\omega)|, & \text{if } |Y_n(j\omega)| < |\bar{Y}_{n-1}(j\omega)| \end{cases},$$
(8.43)

where again β_a is the "attack" coefficient and β_d the "decay" coefficient. To further reduce the spectral fluctuation, both $|\hat{V}_n(j\omega)|$ and $|\bar{Y}_n(j\omega)|$ are averaged across the neighboring frequency bins around ω. Finally, an estimate of the noise spectrum is obtained by multiplying $|\hat{V}_n(j\omega)|/|\bar{Y}_n(j\omega)|$ with $Y_n(j\omega)$, and the time-domain noise signal is obtained through the inverse DFT and the overlap-add technique.

To illustrate the accuracy of the sequential noise estimator, we provide an example where a clean speech signal (same as used in Chapter 7) is corrupted by the car noise with an iSNR = 10 dB. The estimated noise and its spectrogram are plotted in Fig. 8.3. Comparing Fig. 8.3 with Fig. 7.2, one can see that during the absence of speech, the noise estimate is a good approximation of the noise signal. During the presence of speech, the noise estimate consists of some minor speech components. However, such minor amount of speech remained in the noise estimate is almost inaudible.

An apparent advantage of this sequential noise estimation technique is that it does not require an explicit VAD. In addition, such a method is in general able to capture the noise characteristics in both the presence and absence of speech, therefore it works even in the environments where the noise is not very stationary.

8.4.3 Performance Comparison in NYSE Noise

Based upon the previous sequential noise estimation method, we implemented a frequency-domain noise reduction system using the overlap-add technique. Briefly, the input noisy speech is partitioned into overlapped frames with a

90   8 Optimal Filters in the Frequency Domain

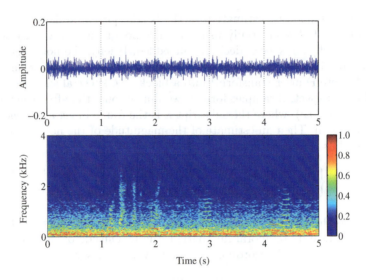

Fig. 8.3. Waveform and spectrogram (first 5 s) of the estimated car noise with an iSNR = 10 dB.

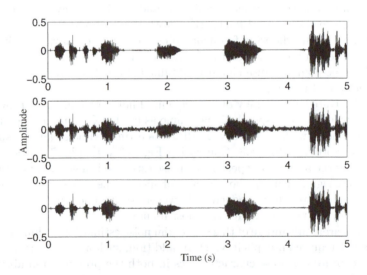

Fig. 8.4. Waveforms of the clean, noisy, and enhanced speech (with the frequency-domain Wiener filter) in NYSE noise with an iSNR = 10 dB.

frame length of 8 ms (corresponding to 64 points at an 8-kHz sampling rate) and an overlap factor of 75%. Each frame is multiplied with a Kaiser window and then transformed into the frequency domain using the FFT. The noise spectrum is then estimated using the sequential noise estimator, and a gain

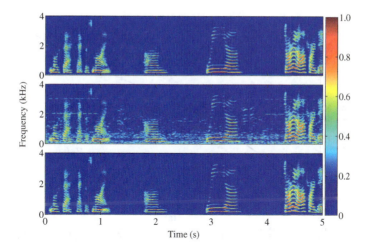

Fig. 8.5. Spectrograms of the clean, noisy, and enhanced speech (with the frequency-domain Wiener filter) in NYSE noise with an iSNR = 10 dB.

filter is then computed according to (8.3), (8.22), (8.23), and (8.33) respectively. The estimated gain filter is subsequently applied to the noisy speech spectrum. The gain-modified noise speech spectrum is then transformed back to the time domain using the inverse FFT (IFFT). A synthesis window, which is the same as the analysis window, is applied, and the result is overlapped with, and added to the previous estimated signal to obtain the final enhanced speech.

Figure 8.4 illustrates the first 5 seconds of the waveforms of the clean speech, noisy speech corrupted by the NYSE noise at an iSNR = 10 dB, and the enhanced speech using the frequency-domain Wiener filter. It shows, at least visibly, that the background noise is significantly reduced using the Wiener filter. To visualize the clean speech estimate in the frequency domain, the spectrograms of the clean, noisy, and enhanced speech signals are displayed in Fig. 8.5. It can be seen that the background noise is dramatically removed, and the spectrogram of the estimated speech resembles that of the original speech.

The detailed evaluation of the output SNR and speech-distortion index as a function of the input SNR in the NYSE noise is sketched in Fig 8.6. [Note that, similar to the previous chapter, we first computed the filtered speech and noise signals $X_F(j\omega)$ and $V_F(j\omega)$, and the output SNR and speech-distortion index are subsequently estimated. This way, the results from different chapters can be compared]. It is noticed that the output SNR is always larger than the input SNR, but the amount of SNR improvement depends on the input SNR. Roughly, the lower the input SNR, the larger the SNR improvement. However, the speech-distortion index tends to increase as more SNR improvement is

92   8 Optimal Filters in the Frequency Domain

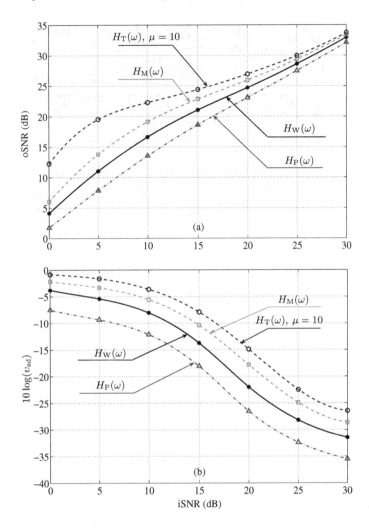

Fig. 8.6. Performance of the different methods in different input SNRs with NYSE noise: (a) output SNR versus input SNR and (b) speech-distortion index.

gained. It is clear that noise reduction is achieved at the price of speech distortion, as we have pointed out before.

Comparatively, the magnitude subtraction method achieves a higher SNR gain as compared to the Wiener filter, but it also has a larger speech-distortion index. The power subtraction is less effective than the Wiener filter in terms of SNR improvement, but it causes less speech distortion. Therefore, these two approaches can be chosen to replace the Wiener filter if a compromise between noise reduction and speech distortion is needed. If more noise reduction is required, we can use the magnitude subtraction. If, on the other hand, we

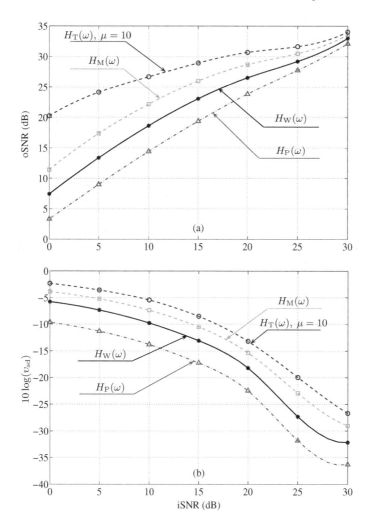

Fig. 8.7. Performance of the different methods in different input SNRs with car noise: (a) output SNR versus input SNR and (b) speech-distortion index.

need to keep the speech distortion low, we can use the power subtraction. Alternatively, the compromise can be achieved through the tradeoff filter by selecting an appropriate value of the parameter μ.

8.4.4 Performance Comparison in Car Noise

Another set of experiments were performed in the car noise condition. The results are plotted in Fig. 8.7. In general, we see that each method has gained more SNR improvement in the car noise than that in the NYSE noise. But

94   8 Optimal Filters in the Frequency Domain

the trend of the SNR improvement and speech-distortion index with respect to the input SNR in the car noise is similar to that in the NYSE noise.

8.5 Summary

In this chapter, we discussed many noise reduction filters in the frequency domain. Both the Wiener and tradeoff filters are optimal in the MMSE sense. The difference between the two is that the former is derived from an unconstrained MSE criterion, while the latter is obtained from a constrained MSE criterion. The magnitude subtraction and power subtraction filters (or more generally the parametric Wiener filter) have no optimality properties associated with them, but they make good sense in practical situations for managing the compromise between noise reduction and speech distortion. We have also discussed many interesting properties of these filters and some of these properties were verified through experiments.

9

Optimal Filters in the KLE Domain

We have seen the advantages of transforming the noisy speech into the frequency domain and then performing noise reduction. Although it is the most studied one, the frequency domain is not the only domain in which we can design interesting filters for noise reduction. There are other domains that may offer more advantages in the filter design. In this chapter, we consider the KLE domain. we derive two broad classes of optimal filters in this domain depending basically on the subband filter length. The first class, similar to the frequency-domain filters, estimates a frame of the clean speech by filtering the corresponding frame of the noisy speech while the second class does noise reduction by filtering not only the current frame but also a number of previous consecutive frames of the noisy speech.

9.1 Class I

In this first category of filters, we consider the particular case where $L_1 = L_2 = \cdots = L_L = 1$. Hence $\mathbf{h}_l = h_{l,0}$, $l = 1, 2, \ldots, L$, are simply scalars. For this class of filters, we always have

$$\text{oSNR}(h_{l,0}) = \text{iSNR}_l, \ \forall l. \tag{9.1}$$

Therefore, the subband output SNR cannot be improved with respect to the subband input SNR. But the fullband output SNR can be improved with respect to the input SNR and from Chapter 3 we know that it is upper bounded (for all filters) as follows:

$$\text{oSNR}(h_{1:L}) \leq \sum_{l=1}^{L} \text{iSNR}_l = \sum_{l=1}^{L} \text{oSNR}(h_{l,0}). \tag{9.2}$$

9.1.1 Wiener Filter

By minimizing $J(h_{l,0})$ [eq. (4.29)] with respect to $h_{l,0}$, we easily find the Wiener filter:

J. Benesty et al., *Noise Reduction in Speech Processing*, Springer Topics in Signal Processing 2,
DOI 10.1007/978-3-642-00296-0_9, © Springer-Verlag Berlin Heidelberg 2009

96   9 Optimal Filters in the KLE Domain

$$h_{W,l,0} = \frac{E\left[c_{x,l}^2(k)\right]}{E\left[c_{y,l}^2(k)\right]}$$

$$= 1 - \frac{E\left[c_{v,l}^2(k)\right]}{E\left[c_{y,l}^2(k)\right]}$$

$$= \frac{\lambda_l}{\lambda_l + \mathbf{q}_l^T \mathbf{R}_{vv} \mathbf{q}_l}$$

$$= \frac{\text{iSNR}_l}{1 + \text{iSNR}_l}. \tag{9.3}$$

This filter is the equivalent form of the frequency-domain Wiener filter (see Chapter 8).

Another way to define the Wiener filter is with the subband SPCCs. Indeed, it is easy to see that

$$h_{W,l,0} = \rho^2(c_{x,l}, c_{y,l})$$
$$= 1 - \rho^2(c_{v,l}, c_{y,l}). \tag{9.4}$$

Since $0 \leq \rho^2(c_{x,l}, c_{y,l}) \leq 1$, therefore $0 \leq h_{W,l,0} \leq 1$. The KLE-domain Wiener filter acts like a gain function. When the level of noise is high in the subband l, $\rho^2(c_{v,l}, c_{y,l}) \approx 1$, then $h_{W,l,0}$ is close to 0 since there is a large amount of noise that has to be removed. When the level of noise is low in the subband l, $\rho^2(c_{v,l}, c_{y,l}) \approx 0$, then $h_{W,l,0}$ is close to 1 and is not going to affect much the signals since there is little noise that needs to be removed.

Now, let us define the complex number

$$\varrho\left(c_{x,l}, c_{v,l}\right) = \rho(c_{x,l}, c_{y,l}) + j\rho(c_{v,l}, c_{y,l})$$
$$= \cos\theta_l + j\sin\theta_l, \tag{9.5}$$

where θ_l is the angle of $\varrho\left(c_{x,l}, c_{v,l}\right)$ for which the modulus is equal to 1. On the complex plane, $\varrho\left(c_{x,l}, c_{v,l}\right)$ is on the unit circle. Since $0 \leq \rho(c_{x,l}, c_{y,l}) \leq 1$ and $0 \leq \rho(c_{v,l}, c_{y,l}) \leq 1$, therefore $0 \leq \theta_l \leq \frac{\pi}{2}$. We can then rewrite the Wiener filter as a function of the angle θ_l:

$$h_{W,l,0} = \cos^2\theta_l$$
$$= 1 - \sin^2\theta_l. \tag{9.6}$$

Hence

$$\lim_{\theta_l \to 0} h_{W,l,0} = 1, \tag{9.7}$$

$$\lim_{\theta_l \to \frac{\pi}{2}} h_{W,l,0} = 0. \tag{9.8}$$

We deduce the subband noise-reduction factor and subband speech-distortion index

$$\xi_{\mathrm{nr}}\left(h_{\mathrm{W},l,0}\right) = \frac{1}{\cos^4\theta_l} \geq 1, \tag{9.9}$$

$$\upsilon_{\mathrm{sd}}\left(h_{\mathrm{W},l,0}\right) = \sin^4\theta_l \leq 1, \tag{9.10}$$

and the subband MNMSE

$$\tilde{J}\left(h_{\mathrm{W},l,0}\right) = h_{\mathrm{W},l,0}, \tag{9.11}$$

which is exactly the Wiener filter. We see clearly how noise reduction and speech distortion depend on the angle θ_l in the KLE-domain Wiener filter. When θ_l increases so does $\xi_{\mathrm{nr}}\left(h_{\mathrm{W},l,0}\right)$; at the same time $\upsilon_{\mathrm{sd}}\left(h_{\mathrm{W},l,0}\right)$ increases. It is also easy to check the formula:

$$\upsilon_{\mathrm{sd}}\left(h_{\mathrm{W},l,0}\right) = 1 - \frac{2}{\sqrt{\xi_{\mathrm{nr}}\left(h_{\mathrm{W},l,0}\right)}} + \frac{1}{\xi_{\mathrm{nr}}\left(h_{\mathrm{W},l,0}\right)}. \tag{9.12}$$

Property 9.1. With the optimal KLE-domain Wiener filter given in (9.3), the fullband output SNR is always greater than or equal to the input SNR, i.e., $\mathrm{oSNR}(h_{\mathrm{W},1:L}) \geq \mathrm{iSNR}$.

Proof. Let us evaluate the fullband SPCC between the two vectors $\mathbf{c}_y(k)$ and $\mathbf{c}_{z,\mathrm{W}}(k) = [h_{\mathrm{W},1,0}c_{y,1}(k) \; h_{\mathrm{W},2,0}c_{y,2}(k) \; \cdots \; h_{\mathrm{W},L,0}c_{y,L}(k)]^T$:

$$
\begin{aligned}
\rho^2\left(\mathbf{c}_y, \mathbf{c}_{z,\mathrm{W}}\right) &= \frac{\left\{\sum_{l=1}^{L} h_{\mathrm{W},l,0}E\left[c_{y,l}^2(k)\right]\right\}^2}{\sum_{l=1}^{L} E\left[c_{y,l}^2(k)\right] \cdot \sum_{l=1}^{L} h_{\mathrm{W},l,0}^2 E\left[c_{y,l}^2(k)\right]} \\
&= \frac{\sum_{l=1}^{L} E\left[c_{x,l}^2(k)\right]}{\sum_{l=1}^{L} E\left[c_{y,l}^2(k)\right]} \cdot \frac{\sum_{l=1}^{L} E\left[c_{y,l}^2(k)\right]}{\sum_{l=1}^{L} h_{\mathrm{W},l,0}E\left[c_{x,l}^2(k)\right]} \\
&= \frac{\rho^2\left(\mathbf{c}_x, \mathbf{c}_y\right)}{\rho^2\left(\mathbf{c}_x, \mathbf{c}_{z,\mathrm{W}}\right)}.
\end{aligned}
$$

Therefore

$$\rho^2\left(\mathbf{c}_x, \mathbf{c}_y\right) = \rho^2\left(\mathbf{c}_y, \mathbf{c}_{z,\mathrm{W}}\right) \cdot \rho^2\left(\mathbf{c}_x, \mathbf{c}_{z,\mathrm{W}}\right) \leq \rho^2\left(\mathbf{c}_x, \mathbf{c}_{z,\mathrm{W}}\right).$$

Using (6.76) and Property 6.32 in the previous expression, we get

$$\frac{\mathrm{iSNR}}{1 + \mathrm{iSNR}} \leq \frac{\mathrm{oSNR}(h_{\mathrm{W},1:L})}{1 + \mathrm{oSNR}(h_{\mathrm{W},1:L})},$$

as a result

$$\mathrm{oSNR}(h_{\mathrm{W},1:L}) \geq \mathrm{iSNR}.$$

98   9 Optimal Filters in the KLE Domain

Substituting (9.3) into (4.36), we find the fullband MNMSE:

$$\tilde{J}(h_{\mathrm{W},1:L}) = 1 - \frac{\sum_{l=1}^{L} \left(\mathbf{q}_l^T \mathbf{R}_v \mathbf{q}_l\right)^2 \left(\lambda_l + \mathbf{q}_l^T \mathbf{R}_v \mathbf{q}_l\right)^{-1}}{\sum_{l=1}^{L} \mathbf{q}_l^T \mathbf{R}_v \mathbf{q}_l} \leq 1. \qquad (9.13)$$

We can compute the fullband speech-distortion index by substituting (9.3) into (3.27):

$$\upsilon_{\mathrm{sd}}(h_{\mathrm{W},1:L}) = 1 - \frac{\mathrm{oSNR}(h_{\mathrm{W},l:L}) + 2}{\mathrm{iSNR} \cdot \xi_{\mathrm{nr}}(h_{\mathrm{W},l:L})} \leq 1. \qquad (9.14)$$

Using (4.36) and (9.14), we get the fullband noise-reduction factor:

$$\xi_{\mathrm{nr}}(h_{\mathrm{W},1:L}) = \frac{\mathrm{oSNR}(h_{\mathrm{W},1:L}) + 1}{\mathrm{iSNR} - \tilde{J}(h_{\mathrm{W},1:L})} \geq 1. \qquad (9.15)$$

Property 9.2. We have

$$\tilde{J}(h_{\mathrm{W},1:L}) = \mathrm{iSNR}\left[1 - \rho^2(\mathbf{c}_x, \mathbf{c}_{z,\mathrm{W}})\right]. \qquad (9.16)$$

Proof. Indeed

$$\begin{aligned} \tilde{J}(h_{\mathrm{W},1:L}) &= \frac{\sum_{l=1}^{L} \lambda_l}{\sum_{l=1}^{L} \mathbf{q}_l^T \mathbf{R}_v \mathbf{q}_l} - \frac{\sum_{l=1}^{L} \lambda_l^2 \left(\lambda_l + \mathbf{q}_l^T \mathbf{R}_v \mathbf{q}_l\right)^{-1}}{\sum_{l=1}^{L} \mathbf{q}_l^T \mathbf{R}_v \mathbf{q}_l} \\ &= \mathrm{iSNR}\left[1 - \frac{\sum_{l=1}^{L} \lambda_l^2 \left(\lambda_l + \mathbf{q}_l^T \mathbf{R}_v \mathbf{q}_l\right)^{-1}}{\sum_{l=1}^{L} \lambda_l}\right] \\ &= \mathrm{iSNR}\left[1 - \rho^2(\mathbf{c}_x, \mathbf{c}_{z,\mathrm{W}})\right]. \end{aligned}$$

Therefore, the fullband NMSE is minimized when the fullband SPCC between the two vectors $\mathbf{c}_x(k)$ and $\mathbf{c}_{z,\mathrm{W}}(k)$ is maximized. This fullband SPCC can be rewritten as follows:

$$\rho^2(\mathbf{c}_x, \mathbf{c}_{z,\mathrm{W}}) = \frac{1}{\xi_{\mathrm{sr}}(h_{\mathrm{W},1:L})} \cdot \frac{1 + \mathrm{oSNR}(h_{\mathrm{W},1:L})}{\mathrm{oSNR}(h_{\mathrm{W},1:L})}. \qquad (9.17)$$

We observe that the KLE-domain Wiener filter is compromising between speech reduction (i.e., speech distortion) and fullband output SNR improvement.

Property 9.3. We have

$$\frac{\mathrm{iSNR}}{1 + \mathrm{oSNR}(h_{\mathrm{W},1:L})} \leq \tilde{J}(h_{\mathrm{W},1:L}) \leq \frac{\mathrm{iSNR}}{1 + \mathrm{iSNR}}. \qquad (9.18)$$

Proof. Since

$$\rho^2(\mathbf{c}_x, \mathbf{c}_{z,\mathrm{W}}) \geq \frac{\mathrm{iSNR}}{1 + \mathrm{iSNR}}$$

and with the help of (9.16), we easily get

$$\tilde{J}(h_{\mathrm{W},1:L}) \le \frac{\mathrm{iSNR}}{1+\mathrm{iSNR}}.$$

Since

$$\rho^2(\mathbf{c}_x, \mathbf{c}_{z,\mathrm{W}}) \le \frac{\mathrm{oSNR}(h_{\mathrm{W},1:L})}{1+\mathrm{oSNR}(h_{\mathrm{W},1:L})}$$

and, again, with the help of (9.16), we obtain

$$\frac{\mathrm{iSNR}}{1+\mathrm{oSNR}(h_{\mathrm{W},1:L})} \le \tilde{J}(h_{\mathrm{W},1:L}).$$

Hence, we get better bounds for the fullband MNMSE than the usual ones $[0 \le \tilde{J}(h_{\mathrm{W},1:L}) \le 1]$.

Property 9.4. We have

$$\frac{[1+\mathrm{oSNR}(h_{\mathrm{W},1:L})]^2}{\mathrm{iSNR}\cdot\mathrm{oSNR}(h_{\mathrm{W},1:L})} \le \xi_{\mathrm{nr}}(h_{\mathrm{W},1:L}) \le \frac{(1+\mathrm{iSNR})\,[1+\mathrm{oSNR}(h_{\mathrm{W},1:L})]}{\mathrm{iSNR}^2}.$$
$$(9.19)$$

Proof. Easy to show by using (9.15) and the bounds of $\tilde{J}(h_{\mathrm{W},1:L})$ [eq. (9.18)].

Property 9.5. We have

$$\frac{1}{[1+\mathrm{oSNR}(h_{\mathrm{W},1:L})]^2} \le \upsilon_{\mathrm{sd}}(h_{\mathrm{W},1:L}) \le \frac{1+\mathrm{oSNR}(h_{\mathrm{W},1:L})-\mathrm{iSNR}}{(1+\mathrm{iSNR})\,[1+\mathrm{oSNR}(h_{\mathrm{W},1:L})]}.$$
$$(9.20)$$

Proof. Easy to show by using (9.14) and the bounds of $\xi_{\mathrm{nr}}(h_{\mathrm{W},1:L})$ [eq. (9.19)].

It is of great interest to understand how the time-domain Wiener filter filter (see Chapter 7)

$$\mathbf{H}_{\mathrm{W}} = \mathbf{Q}\boldsymbol{\Lambda}\left[\boldsymbol{\Lambda} + \mathbf{Q}^T\mathbf{R}_v\mathbf{Q}\right]^{-1}\mathbf{Q}^T \tag{9.21}$$

is related to the KLE-domain Wiener filter given in (9.3).

Substituting the KLE-domain Wiener filter into (2.30), we see that the estimator of the vector $\mathbf{x}(k)$ can be written as

$$\begin{aligned}
\mathbf{z}_{\mathrm{KLE},\mathrm{W}}(k) &= \sum_{l=1}^{L} h_{\mathrm{W},l,0} c_{y,l}\mathbf{q}_l \\
&= \left(\sum_{l=1}^{L} h_{\mathrm{W},l,0}\mathbf{q}_l\mathbf{q}_l^T\right)\mathbf{y}(k) \\
&= \mathbf{H}_{\mathrm{KLE},\mathrm{W}}\mathbf{y}(k).
\end{aligned} \tag{9.22}$$

100   9 Optimal Filters in the KLE Domain

Therefore, the time-domain filter

$$\mathbf{H}_{\mathrm{KLE,W}} = \sum_{l=1}^{L} h_{\mathrm{W},l,0} \mathbf{q}_l \mathbf{q}_l^T \tag{9.23}$$

is strictly equivalent to the KLE-domain filter $h_{\mathrm{W},l,0}$, $l = 1, 2, \ldots, L$. Substituting (9.3) into (9.23), we easily find that

$$\mathbf{H}_{\mathrm{KLE,W}} = \mathbf{Q}\boldsymbol{\Lambda} \left[\boldsymbol{\Lambda} + \mathrm{diag}\left(\mathbf{Q}^T \mathbf{R}_v \mathbf{Q}\right) \right]^{-1} \mathbf{Q}^T. \tag{9.24}$$

Clearly, the two filters $\mathbf{H}_{\mathrm{W}}$ and $\mathbf{H}_{\mathrm{KLE,W}}$ are very close. For example if the noise is white, then $\mathbf{H}_{\mathrm{W}} = \mathbf{H}_{\mathrm{KLE,W}}$. Also the orthogonal matrix $\mathbf{Q}$ tends to diagonalize the Toeplitz matrix $\mathbf{R}_v$. In this case, $\mathbf{Q}^T \mathbf{R}_v \mathbf{Q} \approx \mathrm{diag}\left(\mathbf{Q}^T \mathbf{R}_v \mathbf{Q}\right)$ and as a result, $\mathbf{H}_{\mathrm{W}} \approx \mathbf{H}_{\mathrm{KLE,W}}$.

9.1.2 Parametric Wiener Filter

One practical way to control the compromise between noise reduction and speech distortion is via the parametric Wiener filter [50]:

$$h_{\mathrm{G},l,0} = \left[1 - \sin^{\beta_1} \theta_l \right]^{\beta_2}, \tag{9.25}$$

where β_1 and β_2 are two positive parameters that allow the control of this compromise. For $(\beta_1, \beta_2) = (2, 1)$, we get the KLE-domain Wiener filter developed in the previous subsection. Taking $(\beta_1, \beta_2) = (2, 1/2)$, leads to

$$h_{\mathrm{P},l,0} = \sqrt{1 - \sin^2 \theta_l} \tag{9.26}$$
$$= \cos \theta_l,$$

which is the equivalent form of the power subtraction method. The pair $(\beta_1, \beta_2) = (1, 1)$ gives the equivalent form of the magnitude subtraction method:

$$h_{\mathrm{M},l,0} = 1 - \sin \theta_l \tag{9.27}$$
$$= 1 - \sqrt{1 - \cos^2 \theta_l}.$$

We can verify that the subband noise-reduction factors for the power subtraction and magnitude subtraction methods are

$$\xi_{\mathrm{nr}}\left(h_{\mathrm{P},l,0}\right) = \frac{1}{\cos^2 \theta_l}, \tag{9.28}$$

$$\xi_{\mathrm{nr}}\left(h_{\mathrm{M},l,0}\right) = \frac{1}{(1 - \sin \theta_l)^2}, \tag{9.29}$$

and the corresponding subband speech-distortion indices are

$$v_{\text{sd}}\left(h_{\text{P},l,0}\right) = \left(1 - \cos\theta_l\right)^2, \tag{9.30}$$

$$v_{\text{sd}}\left(h_{\text{M},l,0}\right) = \sin^2\theta_l. \tag{9.31}$$

We can also easily check that

$$\xi_{\text{nr}}\left(h_{\text{M},l,0}\right) \geq \xi_{\text{nr}}\left(h_{\text{W},l,0}\right) \geq \xi_{\text{nr}}\left(h_{\text{P},l,0}\right), \tag{9.32}$$

$$v_{\text{sd}}\left(h_{\text{P},l,0}\right) \leq v_{\text{sd}}\left(h_{\text{W},l,0}\right) \leq v_{\text{sd}}\left(h_{\text{M},l,0}\right). \tag{9.33}$$

The two previous inequalities show that, among the three methods, the magnitude subtraction is the most aggressive one as far as noise reduction is concerned, a very well-known fact in the literature [38], but at the same time it's the one that will likely distorts most the speech signal. The smoothest approach is the power subtraction while the Wiener filter is between the two others in terms of speech distortion and noise reduction.

In the previous subsection, we have shown that it was possible to rewrite the KLE-domain Wiener filter into the time domain. This way, it is more convenient to compare the two Wiener filters from a theoretical point of view (in the time and KLE domains). Following the same line of thoughts, it is easy to show that the KLE-domain power subtraction and magnitude subtraction algorithms are equivalent to

$$\mathbf{H}_{\text{KLE,P}} = \mathbf{Q}\mathbf{\Lambda}^{1/2}\left[\mathbf{\Lambda} + \text{diag}\left(\mathbf{Q}^T\mathbf{R}_v\mathbf{Q}\right)\right]^{-1/2}\mathbf{Q}^T \tag{9.34}$$

and

$$\mathbf{H}_{\text{KLE,M}} = \mathbf{I} - \mathbf{Q}\left[\mathbf{I} + \mathbf{\Lambda}\text{diag}^{-1}\left(\mathbf{Q}^T\mathbf{R}_v\mathbf{Q}\right)\right]^{-1/2}\mathbf{Q}^T. \tag{9.35}$$

9.1.3 Tradeoff Filter

The tradeoff filter is obtained by minimizing the speech distortion with the constraint that the residual noise level is equal to a value smaller than the level of the original noise. This is equivalent to solving the problem

$$\min_{h_{l,0}} J_x(h_{l,0}) \quad \text{subject to} \quad J_v(h_{l,0}) = \beta \cdot \mathbf{q}_l^T\mathbf{R}_v\mathbf{q}_l, \tag{9.36}$$

where

$$J_x(h_{l,0}) = (1 - h_{l,0})^2\,\lambda_l, \tag{9.37}$$

$$J_v(h_{l,0}) = h_{l,0}^2 \cdot \mathbf{q}_l^T\mathbf{R}_v\mathbf{q}_l, \tag{9.38}$$

and $0 < \beta < 1$ in order to have some noise reduction in the subband l. If we use a Lagrange multiplier, $\mu \geq 0$, to adjoin the constraint to the cost function, we get the tradeoff filter:

102   9 Optimal Filters in the KLE Domain

$$
\begin{aligned}
h_{\mathrm{T},l,0} &= \frac{\lambda_l}{\lambda_l + \mu \cdot \mathbf{q}_l^T \mathbf{R}_v \mathbf{q}_l} \\
&= \frac{E\left[c_{y,l}^2(k)\right] - \mathbf{q}_l^T \mathbf{R}_v \mathbf{q}_l}{E\left[c_{y,l}^2(k)\right] + (\mu - 1)\mathbf{q}_l^T \mathbf{R}_v \mathbf{q}_l} \\
&= \frac{\mathrm{iSNR}_l}{\mu + \mathrm{iSNR}_l}.
\end{aligned}
\tag{9.39}
$$

This filter can be seen as a KLE-domain Wiener filter with adjustable input noise level $\mu \cdot \mathbf{q}_l^T \mathbf{R}_v \mathbf{q}_l$.

The subband SPCC between the two signals $c_{x,l}(k)$ and $c_{x,l}(k) + \sqrt{\mu} c_{v,l}(k)$ in the subband l is

$$
\rho^2\left(c_{x,l}, c_{x,l} + \sqrt{\mu} c_{v,l}\right) = \frac{\mathrm{iSNR}_l}{\mu + \mathrm{iSNR}_l}.
\tag{9.40}
$$

The subband SPCC between the two signals $c_{v,l}(k)$ and $c_{x,l}(k) + \sqrt{\mu} c_{v,l}(k)$ in the subband l is

$$
\rho^2\left(c_{v,l}, c_{x,l} + \sqrt{\mu} c_{v,l}\right) = \frac{\mu}{\mu + \mathrm{iSNR}_l}.
\tag{9.41}
$$

Therefore, we can write the tradeoff filter as a function of these two subband SPCCs:

$$
\begin{aligned}
h_{\mathrm{T},l,0} &= \rho^2\left(c_{x,l}, c_{x,l} + \sqrt{\mu} c_{v,l}\right) \\
&= 1 - \rho^2\left(c_{v,l}, c_{x,l} + \sqrt{\mu} c_{v,l}\right).
\end{aligned}
\tag{9.42}
$$

Now, let us define the complex number

$$
\begin{aligned}
\varrho_\mu\left(c_{x,l}, c_{v,l}\right) &= \rho\left(c_{x,l}, c_{x,l} + \sqrt{\mu} c_{v,l}\right) + j\rho\left(c_{v,l}, c_{x,l} + \sqrt{\mu} c_{v,l}\right) \\
&= \cos\theta_l(\mu) + j\sin\theta_l(\mu),
\end{aligned}
\tag{9.43}
$$

where $\theta_l(\mu)$ is the angle of $\varrho_\mu\left(c_{x,l}, c_{v,l}\right)$ for which the modulus is equal to 1. On the complex plane, $\varrho_\mu\left(c_{x,l}, c_{v,l}\right)$ is on the unit circle. Since $0 \leq \rho\left(c_{x,l}, c_{x,l} + \sqrt{\mu} c_{v,l}\right) \leq 1$ and $0 \leq \rho\left(c_{v,l}, c_{x,l} + \sqrt{\mu} c_{v,l}\right) \leq 1$, therefore $0 \leq \theta_l(\mu) \leq \frac{\pi}{2}$. We can then rewrite the tradeoff filter as a function of the angle $\theta_l(\mu)$:

$$
\begin{aligned}
h_{\mathrm{T},l,0} &= \cos^2\theta_l(\mu) \\
&= 1 - \sin^2\theta_l(\mu).
\end{aligned}
\tag{9.44}
$$

We deduce the subband noise-reduction factor and subband speech-distortion index

$$
\xi_{\mathrm{nr}}\left(h_{\mathrm{T},l,0}\right) = \frac{1}{\cos^4\theta_l(\mu)} \geq 1,
\tag{9.45}
$$

$$
\upsilon_{\mathrm{sd}}\left(h_{\mathrm{T},l,0}\right) = \sin^4\theta_l(\mu) \leq 1.
\tag{9.46}
$$

Using Properties 3.2 and 3.3, we see how the fullband noise-reduction factor and fullband speech-distortion index are bounded:

$$1 \le \xi_{\mathrm{nr}}\left(h_{\mathrm{T},1:L}\right) \le \sum_{l=1}^{L} \frac{1}{\cos^4 \theta_l(\mu)}, \tag{9.47}$$

$$0 \le \upsilon_{\mathrm{sd}}\left(h_{\mathrm{T},1:L}\right) \le \min\left[\sum_{l=1}^{L} \sin^4 \theta_l(\mu), 1\right]. \tag{9.48}$$

Property 9.6. With the tradeoff filter given in (9.39), the fullband output SNR is always greater than or equal to the input SNR, i.e., $\mathrm{oSNR}(h_{\mathrm{T},1:L}) \ge \mathrm{iSNR}$.

Proof. The fullband SPCC between the two vectors $\mathbf{c}_x(k)$ and $\mathbf{c}_x(k)+\sqrt{\mu}\mathbf{c}_v(k)$ is

$$\rho^2\left(\mathbf{c}_x, \mathbf{c}_x + \sqrt{\mu}\mathbf{c}_v\right) = \frac{\left(\sum_{l=1}^{L} \lambda_l\right)^2}{\left(\sum_{l=1}^{L} \lambda_l\right)\left(\sum_{l=1}^{L} \lambda_l + \mu \sum_{l=1}^{L} \mathbf{q}_l^T \mathbf{R}_v \mathbf{q}_l\right)}$$

$$= \frac{\mathrm{iSNR}}{\mu + \mathrm{iSNR}}.$$

The fullband SPCC between the two vectors $\mathbf{c}_x(k)$ and $\mathbf{c}_{x,\mathrm{F}}(k) + \sqrt{\mu}\mathbf{c}_{v,\mathrm{F}}(k)$ is

$$\rho^2\left(\mathbf{c}_x, \mathbf{c}_{x,\mathrm{F}} + \sqrt{\mu}\mathbf{c}_{v,\mathrm{F}}\right) =$$

$$\frac{\left(\sum_{l=1}^{L} h_{\mathrm{T},l,0}\lambda_l\right)^2}{\left(\sum_{l=1}^{L} \lambda_l\right)\left(\sum_{l=1}^{L} h_{\mathrm{T},l,0}^2\lambda_l + \mu \sum_{l=1}^{L} h_{\mathrm{T},l,0}^2 \cdot \mathbf{q}_l^T \mathbf{R}_v \mathbf{q}_l\right)}$$

$$= \frac{\sum_{l=1}^{L} h_{\mathrm{T},l,0}\lambda_l}{\sum_{l=1}^{L} \lambda_l}.$$

Another way to write the same fullband SPCC is the following:

$$\rho^2\left(\mathbf{c}_x, \mathbf{c}_{x,\mathrm{F}} + \sqrt{\mu}\mathbf{c}_{v,\mathrm{F}}\right) = \frac{\left(\sum_{l=1}^{L} h_{\mathrm{T},l,0}\lambda_l\right)^2}{\left(\sum_{l=1}^{L} \lambda_l\right)\left(\sum_{l=1}^{L} h_{\mathrm{T},l,0}^2\lambda_l\right)} \cdot \frac{\mathrm{oSNR}\left(h_{\mathrm{T},1:L}\right)}{\mu + \mathrm{oSNR}\left(h_{\mathrm{T},1:L}\right)}$$

$$= \rho^2\left(\mathbf{c}_x, \mathbf{c}_{x,\mathrm{F}}\right) \cdot \rho^2\left(\mathbf{c}_{x,\mathrm{F}}, \mathbf{c}_{x,\mathrm{F}} + \sqrt{\mu}\mathbf{c}_{v,\mathrm{F}}\right)$$

$$\le \frac{\mathrm{oSNR}\left(h_{\mathrm{T},1:L}\right)}{\mu + \mathrm{oSNR}\left(h_{\mathrm{T},1:L}\right)}.$$

Now, let us evaluate the fullband SPCC between the two vectors $\mathbf{c}_x(k) + \sqrt{\mu}\mathbf{c}_v(k)$ and $\mathbf{c}_{x,\mathrm{F}}(k) + \sqrt{\mu}\mathbf{c}_{v,\mathrm{F}}(k)$:

104   9 Optimal Filters in the KLE Domain

$$\rho^2 \left(\mathbf{c}_x + \sqrt{\mu}\mathbf{c}_v, \mathbf{c}_{x,\mathrm{F}} + \sqrt{\mu}\mathbf{c}_{v,\mathrm{F}}\right) = \frac{\sum_{l=1}^{L} \lambda_l}{\sum_{l=1}^{L} \lambda_l + \mu \sum_{l=1}^{L} \mathbf{q}_l^T \mathbf{R}_v \mathbf{q}_l} \times$$

$$\frac{\sum_{l=1}^{L} \lambda_l}{\sum_{l=1}^{L} h_{\mathrm{T},l,0} \lambda_l}$$

$$= \frac{\rho^2 \left(\mathbf{c}_x, \mathbf{c}_x + \sqrt{\mu}\mathbf{c}_v\right)}{\rho^2 \left(\mathbf{c}_x, \mathbf{c}_{x,\mathrm{F}} + \sqrt{\mu}\mathbf{c}_{v,\mathrm{F}}\right)}.$$

Therefore

$$\rho^2 \left(\mathbf{c}_x, \mathbf{c}_x + \sqrt{\mu}\mathbf{c}_v\right) = \frac{\mathrm{iSNR}}{\mu + \mathrm{iSNR}}$$

$$= \rho^2 \left(\mathbf{c}_x + \sqrt{\mu}\mathbf{c}_v, \mathbf{c}_{x,\mathrm{F}} + \sqrt{\mu}\mathbf{c}_{v,\mathrm{F}}\right) \times$$

$$\rho^2 \left(\mathbf{c}_x, \mathbf{c}_{x,\mathrm{F}} + \sqrt{\mu}\mathbf{c}_{v,\mathrm{F}}\right)$$

$$\leq \rho^2 \left(\mathbf{c}_x, \mathbf{c}_{x,\mathrm{F}} + \sqrt{\mu}\mathbf{c}_{v,\mathrm{F}}\right)$$

$$\leq \frac{\mathrm{oSNR}\left(h_{\mathrm{T},1:L}\right)}{\mu + \mathrm{oSNR}\left(h_{\mathrm{T},1:L}\right)}.$$

As a result

$$\mathrm{oSNR}\left(h_{\mathrm{T},1:L}\right) \geq \mathrm{iSNR}.$$

Furthermore, by using Property 3.1, we see how the fullband output SNR for the tradeoff filter is bounded:

$$\mathrm{iSNR} \leq \mathrm{oSNR}\left(h_{\mathrm{T},1:L}\right) \leq \sum_{l=1}^{L} \mathrm{iSNR}_l. \tag{9.49}$$

As we already did for other filters in this class, we can write the KLE-domain tradeoff filter into the time domain. Indeed, substituting (9.39) into (2.30), we find that

$$\mathbf{H}_{\mathrm{KLE,T}} = \mathbf{Q}\mathbf{\Lambda} \left[\mathbf{\Lambda} + \mu \cdot \mathrm{diag}\left(\mathbf{Q}^T \mathbf{R}_v \mathbf{Q}\right)\right]^{-1} \mathbf{Q}^T, \tag{9.50}$$

which is identical to the filter proposed in [103]. This filter can be compared to the time-domain tradeoff filter (see Chapter 7)

$$\mathbf{H}_{\mathrm{T}} = \mathbf{Q}\mathbf{\Lambda} \left[\mathbf{\Lambda} + \mu \cdot \mathbf{Q}^T \mathbf{R}_v \mathbf{Q}\right]^{-1} \mathbf{Q}^T. \tag{9.51}$$

We see that if the noise is white, the two filters are the same.

We can make the tradeoff filter more general by making the factor β dependent on the subband index l, i.e., β_l. By doing so, the control between noise reduction and speech distortion can be more effective since each subband l can be controlled independently of the others. With this consideration, we can easily see that the optimal filter derived from the criterion (9.36) is now

$$h_{\mathrm{T},l,0} = \frac{\mathrm{iSNR}_l}{\mu_l + \mathrm{iSNR}_l}, \qquad (9.52)$$

where μ_l is the subband-dependent Lagrange multiplier. This approach can now provide some noise spectral shaping for masking by the speech signal [49], [68], [69], [75], [120], [121].

9.2 Class II

Although they can improve the fullband SNR, the optimal filters derived in Class I have no impact on the subband SNR. In this section, we consider another category of filters $\mathbf{h}_l$, $l = 1, 2, \ldots, L$, with length L_l where at least one of them has a length $L_l > 1$. In this scenario it is possible to improve both the subband (when $L_l > 1$) and fullband SNRs at the same time.

9.2.1 Wiener Filter

From the KLE-domain MSE [eq. (4.29)], we easily deduce the KLE-domain filter:

$$\begin{aligned}
\mathbf{h}_{\mathrm{W},l} &= \mathbf{R}_{c_y,l}^{-1} \mathbf{R}_{c_x,l} \mathbf{i}_l \\
&= \left(\mathbf{I}_l - \mathbf{R}_{c_y,l}^{-1} \mathbf{R}_{c_v,l} \right) \mathbf{i}_l,
\end{aligned} \qquad (9.53)$$

where $\mathbf{I}_l$ is the identity matrix of size $L_l \times L_l$. Another way to write this filter is

$$\begin{aligned}
\mathbf{h}_{\mathrm{W},l} &= \left(\frac{\mathbf{I}_l}{\mathrm{iSNR}_l} + \tilde{\mathbf{R}}_{c_v,l}^{-1} \tilde{\mathbf{R}}_{c_x,l} \right)^{-1} \tilde{\mathbf{R}}_{c_v,l}^{-1} \tilde{\mathbf{R}}_{c_x,l} \mathbf{i}_l \\
&= \rho^2 \left(c_{x,l}, c_{y,l} \right) \tilde{\mathbf{R}}_{c_y,l}^{-1} \tilde{\mathbf{R}}_{c_x,l} \\
&= \left[\mathbf{I}_l - \rho^2 \left(c_{v,l}, c_{y,l} \right) \tilde{\mathbf{R}}_{c_y,l}^{-1} \tilde{\mathbf{R}}_{c_v,l} \right] \mathbf{i}_l,
\end{aligned} \qquad (9.54)$$

where

$$\tilde{\mathbf{R}}_{c_x,l} = \frac{\mathbf{R}_{c_x,l}}{\lambda_l},$$

$$\tilde{\mathbf{R}}_{c_v,l} = \frac{\mathbf{R}_{c_v,l}}{\mathbf{q}_l^T \mathbf{R}_v \mathbf{q}_l}.$$

We can see from (9.54) that

$$\lim_{\mathrm{iSNR}_l \to \infty} \mathbf{h}_{\mathrm{W},l} = \mathbf{i}_l, \qquad (9.55)$$

$$\lim_{\mathrm{iSNR}_l \to 0} \mathbf{h}_{\mathrm{W},l} = \mathbf{0}_{L_l \times 1}, \qquad (9.56)$$

where $\mathbf{0}_{L_l \times 1}$ is a vector of length L_l with all its elements being zeros.

106   9 Optimal Filters in the KLE Domain

Property 9.7. With the optimal KLE-domain Wiener filter given in (9.53), the subband output SNR is always greater than or equal to the subband input SNR, i.e., $\text{oSNR}(\mathbf{h}_{W,l}) \geq \text{iSNR}_l, \forall l$.

Proof. Let us evaluate the SPCC between $c_{y,l}(k)$ and $\mathbf{h}_{W,l}^T \mathbf{c}_{y,l}(k)$

$$\rho^2\left(c_{y,l}, \mathbf{h}_{W,l}^T \mathbf{c}_{y,l}\right) = \frac{\left(\mathbf{i}_l^T \mathbf{R}_{c_y,l} \mathbf{h}_{W,l}\right)^2}{\left(\lambda_l + \mathbf{q}_l^T \mathbf{R}_v \mathbf{q}_l\right)\left(\mathbf{h}_{W,l}^T \mathbf{R}_{c_y,l} \mathbf{h}_{W,l}\right)}$$

$$= \frac{\lambda_l}{\lambda_l + \mathbf{q}_l^T \mathbf{R}_v \mathbf{q}_l} \cdot \frac{\lambda_l}{\mathbf{i}_l^T \mathbf{R}_{c_x,l} \mathbf{h}_{W,l}}$$

$$= \frac{\rho^2\left(c_{x,l}, c_{y,l}\right)}{\rho^2\left(c_{x,l}, \mathbf{h}_{W,l}^T \mathbf{c}_{y,l}\right)}.$$

Therefore

$$\rho^2\left(c_{x,l}, c_{y,l}\right) = \rho^2\left(c_{y,l}, \mathbf{h}_{W,l}^T \mathbf{c}_{y,l}\right) \cdot \rho^2\left(c_{x,l}, \mathbf{h}_{W,l}^T \mathbf{c}_{y,l}\right) \leq \rho^2\left(c_{x,l}, \mathbf{h}_{W,l}^T \mathbf{c}_{y,l}\right).$$

But

$$\rho^2\left(c_{x,l}, c_{y,l}\right) = \frac{\text{iSNR}_l}{1 + \text{iSNR}_l},$$

$$\rho^2\left(\mathbf{h}_{W,l}^T \mathbf{c}_{x,l}, \mathbf{h}_{W,l}^T \mathbf{c}_{y,l}\right) = \frac{\text{oSNR}(\mathbf{h}_{W,l})}{1 + \text{oSNR}(\mathbf{h}_{W,l})},$$

and

$$\rho^2\left(c_{x,l}, \mathbf{h}_{W,l}^T \mathbf{c}_{y,l}\right) = \rho^2\left(c_{x,l}, \mathbf{h}_{W,l}^T \mathbf{c}_{x,l}\right) \cdot \rho^2\left(\mathbf{h}_{W,l}^T \mathbf{c}_{x,l}, \mathbf{h}_{W,l}^T \mathbf{c}_{y,l}\right)$$

$$\leq \rho^2\left(\mathbf{h}_{W,l}^T \mathbf{c}_{x,l}, \mathbf{h}_{W,l}^T \mathbf{c}_{y,l}\right).$$

Hence

$$\frac{\text{iSNR}_l}{1 + \text{iSNR}_l} \leq \frac{\text{oSNR}(\mathbf{h}_{W,l})}{1 + \text{oSNR}(\mathbf{h}_{W,l})}.$$

As a result

$$\text{oSNR}(\mathbf{h}_{W,l}) \geq \text{iSNR}_l, \forall l.$$

The KLE-domain minimum MSE (MMSE) and KLE-domain minimum NMSE (MNMSE) are obtained by replacing $\mathbf{h}_{W,l}$ in (4.29) and (4.32):

$$J(\mathbf{h}_{W,l}) = \lambda_l - \mathbf{i}_l^T \mathbf{R}_{c_x,l} \mathbf{R}_{c_y,l}^{-1} \mathbf{R}_{c_x,l} \mathbf{i}_l$$

$$= \mathbf{q}_l^T \mathbf{R}_v \mathbf{q}_l - \mathbf{i}_l^T \mathbf{R}_{c_v,l} \mathbf{R}_{c_y,l}^{-1} \mathbf{R}_{c_v,l} \mathbf{i}_l, \tag{9.57}$$

$$\tilde{J}(\mathbf{h}_{W,l}) = 1 - \frac{\mathbf{i}_l^T \mathbf{R}_{c_v,l} \mathbf{R}_{c_y,l}^{-1} \mathbf{R}_{c_v,l} \mathbf{i}_l}{\mathbf{q}_l^T \mathbf{R}_v \mathbf{q}_l} \leq 1. \tag{9.58}$$

We can compute the subband speech-distortion index by substituting (9.53) into (3.25):

$$\upsilon_{sd}\left(\mathbf{h}_{W,l}\right) = 1 - \frac{\text{oSNR}(\mathbf{h}_{W,l}) + 2}{\text{iSNR}_l \cdot \xi_{nr}\left(\mathbf{h}_{W,l}\right)} \le 1. \tag{9.59}$$

Using (4.32) and (9.59), we get the subband noise-reduction factor:

$$\xi_{nr}\left(\mathbf{h}_{W,l}\right) = \frac{\text{oSNR}(\mathbf{h}_{W,l}) + 1}{\text{iSNR}_l - \tilde{J}\left(\mathbf{h}_{W,l}\right)} \ge 1. \tag{9.60}$$

Property 9.8. We have

$$\tilde{J}\left(\mathbf{h}_{W,l}\right) = \text{iSNR}_l \left[1 - \rho^2\left(c_{x,l}, \mathbf{h}_{W,l}^T \mathbf{c}_{y,l}\right)\right]. \tag{9.61}$$

Proof. Indeed

$$\begin{aligned}
\tilde{J}\left(\mathbf{h}_{W,l}\right) &= \frac{\lambda_l}{\mathbf{q}_l^T \mathbf{R}_v \mathbf{q}_l} - \frac{\mathbf{i}_l^T \mathbf{R}_{c_x,l} \mathbf{R}_{c_y,l}^{-1} \mathbf{R}_{c_x,l} \mathbf{i}_l}{\mathbf{q}_l^T \mathbf{R}_v \mathbf{q}_l} \\
&= \text{iSNR}_l \left(1 - \frac{\mathbf{i}_l^T \mathbf{R}_{c_x,l} \mathbf{h}_{W,l}}{\lambda_l}\right) \\
&= \text{iSNR}_l \left[1 - \rho^2\left(c_{x,l}, \mathbf{h}_{W,l}^T \mathbf{c}_{y,l}\right)\right].
\end{aligned}$$

Therefore, the subband NMSE is minimized when the subband SPCC between the two variables $c_{x,l}(k)$ and $\mathbf{h}_{W,l}^T \mathbf{c}_{y,l}(k)$ is maximized. This subband SPCC can be rewritten as follows:

$$\rho^2\left(c_{x,l}, \mathbf{h}_{W,l}^T \mathbf{c}_{y,l}\right) = \frac{1}{\xi_{sr}\left(\mathbf{h}_{W,l}\right)} \cdot \frac{1 + \text{oSNR}(\mathbf{h}_{W,l})}{\text{oSNR}(\mathbf{h}_{W,l})}. \tag{9.62}$$

We observe that the KLE-domain Wiener filter is compromising between subband speech reduction (i.e., subband speech distortion) and subband output SNR improvement.

Property 9.9. We have

$$\frac{\text{iSNR}_l}{1 + \text{oSNR}(\mathbf{h}_{W,l})} \le \tilde{J}\left(\mathbf{h}_{W,l}\right) \le \frac{\text{iSNR}_l}{1 + \text{iSNR}_l}. \tag{9.63}$$

Proof. Since

$$\rho^2\left(c_{x,l}, \mathbf{h}_{W,l}^T \mathbf{c}_{y,l}\right) \ge \frac{\text{iSNR}_l}{1 + \text{iSNR}_l}$$

and with the help of (9.61), we easily get

$$\tilde{J}\left(\mathbf{h}_{W,l}\right) \le \frac{\text{iSNR}_l}{1 + \text{iSNR}_l}.$$

108   9 Optimal Filters in the KLE Domain

Since

$$\rho^2\left(c_{x,l}, \mathbf{h}_{\mathrm{W},l}^T \mathbf{c}_{y,l}\right) \leq \frac{\mathrm{oSNR}(\mathbf{h}_{\mathrm{W},l})}{1 + \mathrm{oSNR}(\mathbf{h}_{\mathrm{W},l})}$$

and, again, with the help of (9.61), we obtain

$$\frac{\mathrm{iSNR}_l}{1 + \mathrm{oSNR}(\mathbf{h}_{\mathrm{W},l})} \leq \tilde{J}(\mathbf{h}_{\mathrm{W},l}).$$

Hence, we get better bounds for the KLE-domain MNMSE than the usual ones $[0 \leq \tilde{J}(\mathbf{h}_{\mathrm{W},l}) \leq 1]$.

Property 9.10. We have

$$\frac{[1 + \mathrm{oSNR}(\mathbf{h}_{\mathrm{W},l})]^2}{\mathrm{iSNR}_l \cdot \mathrm{oSNR}(\mathbf{h}_{\mathrm{W},l})} \leq \xi_{\mathrm{nr}}(\mathbf{h}_{\mathrm{W},l}) \leq \frac{(1 + \mathrm{iSNR}_l)\,[1 + \mathrm{oSNR}(\mathbf{h}_{\mathrm{W},l})]}{\mathrm{iSNR}_l^2}. \tag{9.64}$$

Proof. Easy to show by using (9.60) and the bounds of $\tilde{J}(\mathbf{h}_{\mathrm{W},l})$ [eq. (9.63)].

Property 9.11. We have

$$\frac{1}{[1 + \mathrm{oSNR}(\mathbf{h}_{\mathrm{W},l})]^2} \leq \upsilon_{\mathrm{sd}}(\mathbf{h}_{\mathrm{W},l}) \leq \frac{1 + \mathrm{oSNR}(\mathbf{h}_{\mathrm{W},l}) - \mathrm{iSNR}_l}{(1 + \mathrm{iSNR}_l)\,[1 + \mathrm{oSNR}(\mathbf{h}_{\mathrm{W},l})]}. \tag{9.65}$$

Proof. Easy to show by using (9.59) and the bounds of $\xi_{\mathrm{nr}}(\mathbf{h}_{\mathrm{W},l})$ [eq. (9.64)].

Property 9.12. With the optimal KLE-domain Wiener filter given in (9.53), the fullband output SNR is always greater than or equal to the input SNR, i.e., $\mathrm{oSNR}(\mathbf{h}_{\mathrm{W},1:L}) \geq \mathrm{iSNR}$.

Proof. The proof is identical to the one given after Property 9.1.

It is remarkable that the KLE-domain filter for $L_l > 1$ can improve both the subband and fullband SNRs. As a consequence, we should expect more noise reduction with this filter than the Wiener filters in the time and frequency domains.

Properties 9.2, 9.3, 9.4, and 9.5 apply here as well.

9.2.2 Tradeoff Filter

The tradeoff filter is obtained by solving the following optimization problem:

$$\min_{\mathbf{h}_l} J_x(\mathbf{h}_l) \quad \text{subject to} \quad J_v(\mathbf{h}_l) = \beta \cdot \mathbf{q}_l^T \mathbf{R}_v \mathbf{q}_l, \tag{9.66}$$

where

$$J_x(\mathbf{h}_l) = E\left\{\left[(\mathbf{i}_l - \mathbf{h}_l)^T \mathbf{c}_{x,l}\right]^2\right\}, \tag{9.67}$$

$$J_v(\mathbf{h}_l) = E\left[\left(\mathbf{h}_l^T \mathbf{c}_{v,l}\right)^2\right], \tag{9.68}$$

and $0 < \beta < 1$ in order to have some noise reduction in the subband l. If we use a Lagrange multiplier, $\mu(\geq 0)$, to adjoin the constraint to the cost function, we easily find the optimal filter:

$$\begin{aligned}
\mathbf{h}_{\mathrm{T},l} &= \left(\mathbf{R}_{c_x,l} + \mu \mathbf{R}_{c_v,l}\right)^{-1} \mathbf{R}_{c_x,l} \mathbf{i}_l \\
&= \left[\mathbf{R}_{c_y,l} + (\mu - 1)\mathbf{R}_{c_v,l}\right]^{-1} \left(\mathbf{R}_{c_y,l} - \mathbf{R}_{c_v,l}\right) \mathbf{i}_l \\
&= \left[\mathbf{I}_l + (\mu - 1)\mathbf{R}_{c_y,l}^{-1}\mathbf{R}_{c_v,l}\right]^{-1} \mathbf{h}_{\mathrm{W},l}, \tag{9.69}
\end{aligned}$$

where the Lagrange multiplier satisfies $J_v(\mathbf{h}_l) = \beta \cdot \mathbf{q}_l^T \mathbf{R}_v \mathbf{q}_l$. In practice it's not easy to determine μ. Therefore, when this parameter is chosen in an ad-hoc way, we can see that for

- $\mu = 1$, $\mathbf{h}_{\mathrm{T},l} = \mathbf{h}_{\mathrm{W},l}$; so the tradeoff and Wiener filters are identical;
- $\mu = 0$, $\mathbf{h}_{\mathrm{T},l} = \mathbf{i}_l$; therefore, with this filter there is neither noise reduction nor speech distortion;
- $\mu > 1$, corresponds to more aggressive noise reduction compared to Wiener, so the residual noise level would be lower, but it is achieved at the expense of higher speech distortion;
- $\mu < 1$, corresponds to less aggressive noise compared to Wiener, in this situation we get less speech distortion but less noise reduction as well.

Property 9.13. With the KLE-domain tradeoff filter given in (9.69), the subband output SNR is always greater than or equal to the subband input SNR, i.e., $\mathrm{oSNR}(\mathbf{h}_{\mathrm{T},l}) \geq \mathrm{iSNR}_l, \forall l$.

Proof. The subband SPCC between the two signals $c_{x,l}(k)$ and $\mathbf{h}_{\mathrm{T},l}^T \mathbf{c}_{x,l}(k) + \sqrt{\mu}\mathbf{h}_{\mathrm{T},l}^T \mathbf{c}_{v,l}(k)$ in the subband l is

$$\begin{aligned}
\rho^2\left(c_{x,l}, \mathbf{h}_{\mathrm{T},l}^T \mathbf{c}_{x,l} + \sqrt{\mu}\mathbf{h}_{\mathrm{T},l}^T \mathbf{c}_{v,l}\right) &= \frac{\left(\mathbf{i}_l^T \mathbf{R}_{c_x,l} \mathbf{h}_{\mathrm{T},l}\right)^2}{\lambda_l \left(\mathbf{h}_{\mathrm{T},l}^T \mathbf{R}_{c_x,l} \mathbf{h}_{\mathrm{T},l} + \mu \mathbf{h}_{\mathrm{T},l}^T \mathbf{R}_{c_v,l} \mathbf{h}_{\mathrm{T},l}\right)} \\
&= \frac{\mathbf{i}_l^T \mathbf{R}_{c_x,l} \mathbf{h}_{\mathrm{T},l}}{\lambda_l}.
\end{aligned}$$

Another way to write the same subband SPCC is the following:

110   9 Optimal Filters in the KLE Domain

$$\rho^2\left(c_{x,l}, \mathbf{h}_{\mathrm{T},l}^T \mathbf{c}_{x,l} + \sqrt{\mu}\mathbf{h}_{\mathrm{T},l}^T \mathbf{c}_{v,l}\right) = \frac{\left(\mathbf{i}_l^T \mathbf{R}_{c_x,l}\mathbf{h}_{\mathrm{T},l}\right)^2}{\lambda_l \cdot \mathbf{h}_{\mathrm{T},l}^T \mathbf{R}_{c_x,l}\mathbf{h}_{\mathrm{T},l}} \cdot \frac{\mathrm{oSNR}\left(\mathbf{h}_{\mathrm{T},l}\right)}{\mu + \mathrm{oSNR}\left(\mathbf{h}_{\mathrm{T},l}\right)}$$

$$= \rho^2\left(c_{x,l}, \mathbf{h}_{\mathrm{T},l}^T \mathbf{c}_{x,l}\right) \times$$

$$\rho^2\left(\mathbf{h}_{\mathrm{T},l}^T \mathbf{c}_{x,l}, \mathbf{h}_{\mathrm{T},l}^T \mathbf{c}_{x,l} + \sqrt{\mu}\mathbf{h}_{\mathrm{T},l}^T \mathbf{c}_{v,l}\right)$$

$$\leq \frac{\mathrm{oSNR}\left(\mathbf{h}_{\mathrm{T},l}\right)}{\mu + \mathrm{oSNR}\left(\mathbf{h}_{\mathrm{T},l}\right)}.$$

Now let us evaluate the subband SPCC between the two signals $c_{x,l}(k) + \sqrt{\mu}c_{v,l}(k)$ and $\mathbf{h}_{\mathrm{T},l}^T \mathbf{c}_{x,l}(k) + \sqrt{\mu}\mathbf{h}_{\mathrm{T},l}^T \mathbf{c}_{v,l}(k)$ in the subband l:

$$\rho^2\left(c_{x,l} + \sqrt{\mu}c_{v,l}, \mathbf{h}_{\mathrm{T},l}^T \mathbf{c}_{x,l} + \sqrt{\mu}\mathbf{h}_{\mathrm{T},l}^T \mathbf{c}_{v,l}\right) =$$

$$\frac{\left[\mathbf{i}_l^T\left(\mathbf{R}_{c_x,l} + \mu\mathbf{R}_{c_v,l}\right)\mathbf{h}_{\mathrm{T},l}\right]^2}{\left(\lambda_l + \mu \cdot \mathbf{q}_l^T \mathbf{R}_v \mathbf{q}_l\right)\left[\mathbf{h}_{\mathrm{T},l}^T\left(\mathbf{R}_{c_x,l} + \mu\mathbf{R}_{c_v,l}\right)\mathbf{h}_{\mathrm{T},l}\right]}$$

$$= \frac{\lambda_l}{\lambda_l + \mu \cdot \mathbf{q}_l^T \mathbf{R}_v \mathbf{q}_l} \cdot \frac{\lambda_l}{\mathbf{i}_l^T \mathbf{R}_{c_x,l}\mathbf{h}_{\mathrm{T},l}}$$

$$= \frac{\rho^2\left(c_{x,l}, c_{x,l} + \sqrt{\mu}c_{v,l}\right)}{\rho^2\left(c_{x,l}, \mathbf{h}_{\mathrm{T},l}^T \mathbf{c}_{x,l} + \sqrt{\mu}\mathbf{h}_{\mathrm{T},l}^T \mathbf{c}_{v,l}\right)}.$$

Therefore

$$\rho^2\left(c_{x,l}, c_{x,l} + \sqrt{\mu}c_{v,l}\right) = \frac{\mathrm{iSNR}_l}{\mu + \mathrm{iSNR}_l}$$

$$= \rho^2\left(c_{x,l} + \sqrt{\mu}c_{v,l}, \mathbf{h}_{\mathrm{T},l}^T \mathbf{c}_{x,l} + \sqrt{\mu}\mathbf{h}_{\mathrm{T},l}^T \mathbf{c}_{v,l}\right) \times$$

$$\rho^2\left(c_{x,l}, \mathbf{h}_{\mathrm{T},l}^T \mathbf{c}_{x,l} + \sqrt{\mu}\mathbf{h}_{\mathrm{T},l}^T \mathbf{c}_{v,l}\right)$$

$$\leq \rho^2\left(c_{x,l}, \mathbf{h}_{\mathrm{T},l}^T \mathbf{c}_{x,l} + \sqrt{\mu}\mathbf{h}_{\mathrm{T},l}^T \mathbf{c}_{v,l}\right)$$

$$\leq \frac{\mathrm{oSNR}\left(\mathbf{h}_{\mathrm{T},l}\right)}{\mu + \mathrm{oSNR}\left(\mathbf{h}_{\mathrm{T},l}\right)}.$$

As a result

$$\mathrm{oSNR}\left(\mathbf{h}_{\mathrm{T},l}\right) \geq \mathrm{iSNR}_l, \ \forall l.$$

Property 9.14. With the KLE-domain tradeoff filter given in (9.69), the fullband output SNR is always greater than or equal to the input SNR, i.e., $\mathrm{oSNR}(\mathbf{h}_{\mathrm{T},1:L}) \geq \mathrm{iSNR}$.

Proof. The proof is identical to the one given after Property 9.6.

9.3 Experiments   111

From the two previous properties, we see that the tradeoff filter in the KLE domain has the potential to improve both the subband and fullband SNRs.

It is straightforward to get a more general form of the tradeoff filter by just making the factor β dependent on the subband index l. We then obtain

$$\mathbf{h}_{\mathrm{T},l} = \left(\mathbf{R}_{c_x,l} + \mu_l \mathbf{R}_{c_v,l}\right)^{-1} \mathbf{R}_{c_x,l} \mathbf{i}_l, \qquad (9.70)$$

where now the Lagrange multiplier depends also on the subband index l. This filter could be useful in practice if we want to give a particular shape to the residual noise.

9.3 Experiments

In this section, we study the performance of the two classes of noise reduction filters in the KLE domain through experiments. The experimental setup is the same as used in Chapter 7.

9.3.1 Impact of Forgetting Factor on Performance of Class-I Filters

To implement the Class-I filters developed in Section 9.1, we need to know the correlation matrices $\mathbf{R}_y$ and $\mathbf{R}_v$. The matrix $\mathbf{R}_y$ can be directly estimated from the noisy signal using the recursive approach given in (7.44). Therefore, the forgetting factor α_y plays a critical role in the estimation accuracy of the correlation matrix $\mathbf{R}_y$, which in turn may significantly affect the noise reduction performance. As explained in Chapter 7, the forgetting factor α_y cannot be too large. If it is too large (close to 1), the recursive estimate will essentially be a long-term average and will not be able to follow the short-term variations of the speech signal. As a result, the nature of the speech signal cannot be fully taken advantage of, which limits the noise reduction performance. Conversely, if α_y is too small, the estimation variance of $\mathbf{R}_y(k)$ will be large, which, again, may lead to performance degradation in noise reduction. Furthermore, $\mathbf{R}_y(k)$ may tend to be rank deficient, causing numerical stability problems. Therefore, a proper value of the forgetting factor is very important. In this experiment, we attempt to find the optimal forgetting factor by directly examining the noise reduction performance. White noise is used in this experiment with an iSNR = 10 dB. Similar to that in Section 7.4, we tend not to use any noise estimator, but compute the correlation matrix $\mathbf{R}_v$ directly from the noise signal using a long-term sample average.

Figure 9.1 plots both the output SNR and speech-distortion index as a function of α_y. It is seen that, for all the investigated algorithms, both the output SNR and speech-distortion index bear a nonmonotonic relationship with α_y. Specifically, the output SNR first increases as α_y increases and then

9 Optimal Filters in the KLE Domain

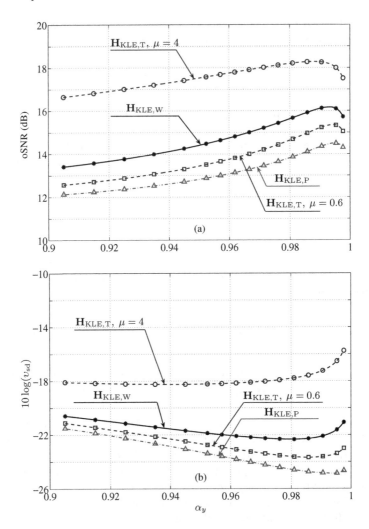

Fig. 9.1. Performance versus the forgetting factor α_y in white Gaussian noise with an iSNR = 10 dB and $L = 20$.

decreases, but the speech-distortion index first decreases with α_y and then increases. The optimal noise reduction performance (highest output SNR and lowest speech distortion) appears when α_y is in the range between 0.985 and 0.995. This result is similar to what was observed with the time-domain filters in Chapter 7.

It is also seen from Fig. 9.1 that the power subtraction method yielded the least SNR gain, but it also has the lowest speech distortion as compared to the Wiener filter and the tradeoff filter with $\mu = 4$ and $\mu = 0.6$. The performance of the tradeoff filter depends on the value of μ. When $\mu = 4$, this method

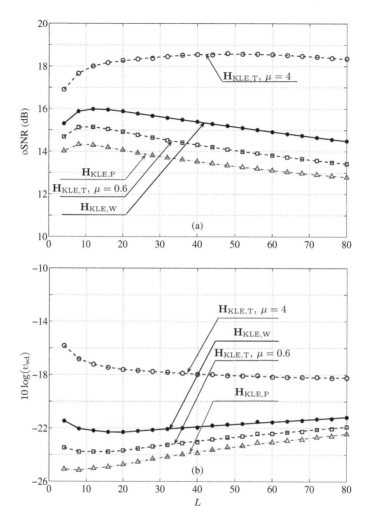

Fig. 9.2. Performance versus filter length L in white Gaussian noise with an iSNR = 10 dB and $\alpha_y = 0.985$.

achieved higher output SNRs than the Wiener filter, but at the cost of higher speech distortion as shown in Fig. 9.1(b). When $\mu = 0.6$, the tradeoff filter yielded less SNR improvement as compared to the Wiener. All these agreed very well with the theoretical analysis given in the previous sections.

9.3.2 Effect of Filter Length on Performance of Class-I Filters

Another important parameter that affects the performance of all the Class-I filters is the filter length L. So, in this experiment, we study the impact of

114   9 Optimal Filters in the KLE Domain

the filter length (also the frame size) L on the performance of noise reduction. Again, white noise is used with an iSNR = 10 dB and the noise correlation matrix is directly computed from the noise signal using a long-term average. Based on the previous experiment, we set $\alpha_y = 0.985$. Figure 9.2 depicts the results. It is safe to say that the length L should be reasonably large enough to achieve good noise reduction performance. When L increases from 1 to 20, the output SNR improves while speech distortion decreases. But if we continue to increase L, there is either marginal additional SNR improvement (for the tradeoff method with $\mu = 4$), or even SNR degradation (for the Wiener filter, the power subtraction, and the tradeoff filter with $\mu = 0.6$), and there is also some increase in speech distortion. In general, good performance for all the studied algorithms is achieved when the filter length L is around 20. This result coincides with what was observed with the time-domain filters in Chapter 7.

9.3.3 Estimation of Clean Speech Correlation Matrix

Previously, we have used the recursive approach to compute the noisy signal correlation matrix $\mathbf{R}_y$. When noise is not stationary, we also used the recursive approach to estimate the correlation matrix $\mathbf{R}_v$. We then assumed that at time k, the clean speech correlation matrix $\mathbf{R}_x$ can be calculated as

$$\mathbf{R}_x(k) = \mathbf{R}_y(k) - \mathbf{R}_v(k). \tag{9.71}$$

Substituting (7.44) and (7.45) into (9.71), one can readily check that $\mathbf{R}_x(k)$ is not in a recursive form (if $\alpha_y \neq \alpha_v$) even though both $\mathbf{R}_y(k)$ and $\mathbf{R}_v(k)$ are computed recursively.

Now we consider to use the recursive method to compute $\mathbf{R}_x$, i.e.,

$$\mathbf{R}_x(k) = \alpha_x \mathbf{R}_x(k-1) + (1 - \alpha_x)\mathbf{x}(k)\mathbf{x}^T(k), \tag{9.72}$$

where α_x is a forgetting factor. Putting (9.72) and (7.45) together and assuming that at any time instant k the relationship (9.71) should be satisfied, we have

$$\mathbf{R}_y(k) = \alpha_x \mathbf{R}_y(k-1) + (1 - \alpha_x)\mathbf{y}(k)\mathbf{y}^T(k) +$$
$$(\alpha_v - \alpha_x)\mathbf{R}_v(k-1) - (\alpha_v - \alpha_x)\mathbf{v}(k)\mathbf{v}^T(k). \tag{9.73}$$

To examine the impact of α_x on the Class-I filters, we repeated the experiment of Section 9.3.1. The only difference is that now the correlation matrix $\mathbf{R}_y(k)$ is computed according to (9.73) and the matrix $\mathbf{R}_v$ is also computed recursively using (7.45) with $\alpha_v = 0.995$. The results are plotted in Fig. 9.3. It is seen from Fig. 9.3 that the output SNR decreases monotonically as the forgetting factor α_x increases; but it decreases slowly at first and then more rapidly. This result is different from what was observed from Fig. 9.1. In comparison, the speech-distortion index bears a nonmonotonic relationship with

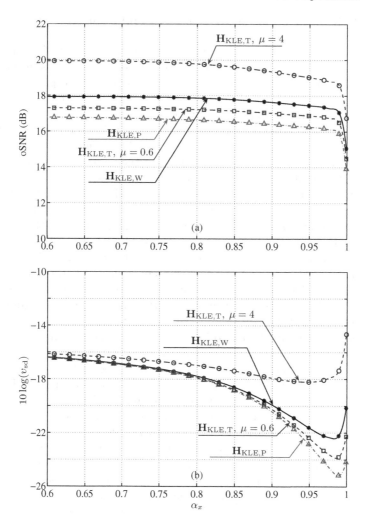

Fig. 9.3. Performance versus the forgetting factor α_x in white Gaussian noise with an iSNR = 10 dB, $\alpha_v = 0.995$, and $L = 20$.

α_x. It first decreases with α_x and then increases. Again, it can be seen that the value of α_x plays a very important role on the noise reduction performance of the studied filters. Comparing Figs. 9.3 and 9.1, one can see that the estimator in (9.73) produces a more consistent noise reduction performance (with respect to the value of the forgetting factor) than the recursive estimator given in (7.44). This is more true in the subband case where, if we need to estimate the correlation matrices, the forgetting factors are much smaller. We will come back to this point when we will discuss the Class-II filters.

116   9 Optimal Filters in the KLE Domain

9.3.4 Performance of Class-I Filters in Different Noise Conditions

In this experiment, we test the performance of the Class-I filters in different input SNRs and noise conditions. We consider three types of noise (see Chapter 7): white Gaussian, car, and NYSE. Based on the previous experiment, we set $L = 20$ and compute the matrix $\mathbf{R}_y(k)$ according to (9.73) with $\alpha_x = 0.985$. Again, we do not use any noise estimator, but compute the noise correlation matrix directly from the noise signal using the recursive method with $\alpha_v = 0.995$ for all the three noise signals. The results of this experiment are shown in Fig. 9.4, where we only plotted the results of the Wiener filter and the tradeoff filter with $\mu = 4$ to simplify the presentation.

It is noticed from Fig. 9.4 that the output SNR is always larger than the input SNR, but the amount of SNR improvement depends on the input SNR and noise characteristics. Roughly, the lower the input SNR, the larger the SNR improvement. However, the speech-distortion index tends to increase as more SNR improvement is gained. This result coincides with what was observed with the time- and frequency-domain filters. If the input SNR is very low (e.g., below 0 dB), the noise reduction filters may even lead to speech quality degradation (instead of improving the speech quality) because of the significant speech distortion. Again, we suggest to use the grace-degradation technique in practical systems to better control the performance.

It is also seen from Fig. 9.4 that both the Wiener and tradeoff filters are less effective in suppressing the NYSE noise as compared to the white noise and the car noise. The main reason for this is due to the fact that the NYSE noise is highly nonstationary. Another way to look at this problem is by examining the eigenvalue distribution of the correlation matrices of the different noise signals. We computed the correlation matrices (with $L = 20$) of the clean speech and all the three noise signals using the long-term average. The diagonalization process given in (2.16) was then applied to find the eigenvalues of all these matrices. As shown in Fig. 9.5, the white noise has constant eigenvalues. The car noise has only a couple of large eigenvalues and all the others tend to be close to 0. Comparatively, the NYSE noise has a similar eigenvalue distribution to the clean speech. These eigenvalue distribution patterns indicate that the NYSE noise is more difficult to detect making the noise suppression more challenging.

9.3.5 Impact of Forgetting Factor on Performance of Class-II Filters

Unlike the Class-I filters where each frame may have a different transformation $\mathbf{Q}$, the Class-II algorithms assume that all the frames share the same transformation $\mathbf{Q}$ (otherwise, filtering the KLE coefficients across different frames would not make much sense). In this situation, the estimation of $\mathbf{Q}$ is relatively easier than that for the Class-I filters. We can simply use a long-term sample average to compute the correlation matrices $\mathbf{R}_y$ and $\mathbf{R}_v$, thereby obtaining

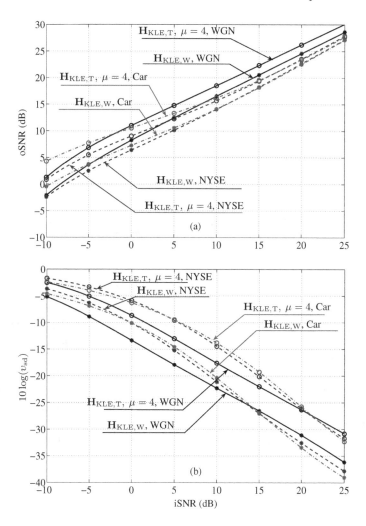

Fig. 9.4. Performance as a function of the input SNR with white Gaussian noise (WGN), car noise, and NYSE babbling noise.

an estimate of $\mathbf{R}_x$. The KLT matrix $\mathbf{Q}$ can then be computed using the eigenvalue decomposition. In our study, we found that the estimation accuracy of the matrix $\mathbf{Q}$ plays a less important role in noise reduction performance of the Class-II methods than it does in performance of the Class-I filters. We can even replace the matrix $\mathbf{Q}$ by the Fourier matrix without degrading the noise reduction performance, which indicates that the idea of the Class-II filters can also be used in the frequency-domain approaches. However, following the theoretical development in Section 9.2, we still use the transformation matrix $\mathbf{Q}$ in our experiments, with the correlation matrices $\mathbf{R}_y$ and $\mathbf{R}_v$ being estimated

118   9 Optimal Filters in the KLE Domain

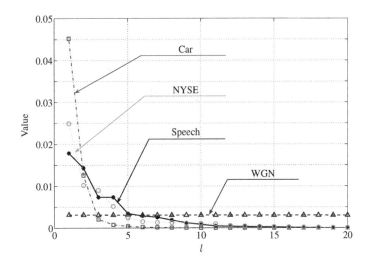

Fig. 9.5. Eigenvalues of the clean speech and different noise signals with $L = 20$.

using a long-term average and the matrix $\mathbf{R}_x$ being computed as $\mathbf{R}_y - \mathbf{R}_v$. The matrix $\mathbf{Q}$ is then applied to each frame of the signals to compute the KLE coefficients $c_{y,l}$ and $c_{v,l}$.

In order to estimate $\mathbf{R}_{c_v,l}$, we need to have an estimate of the noise signal $v(k)$. Although we have developed a noise detector in Chapter 8, we compute the noise statistics directly from the noise signal in this experiment to avoid the influence of the noise estimation error on the parameter optimization. Specifically, the KLT $\mathbf{Q}$ is applied to the noise signal $\mathbf{v}(k)$ to obtain the KLE coefficients $c_{v,l}(k)$. The matrix $\mathbf{R}_{c_v,l}$ is then computed using the recursive method

$$\mathbf{R}_{c_v,l}(k) = \alpha_{c_v,l}\mathbf{R}_{c_v,l}(k-1) + (1 - \alpha_{c_v,l})\mathbf{c}_{v,l}(k)\mathbf{c}_{v,l}^T(k), \qquad (9.74)$$

where $\alpha_{c_v,l}$, same as α_v and α_y, is a forgetting factor.

In the previous analysis, we have pointed out that when the forgetting factors are small, which is the case in the Class-II filters, it is better to assume that the speech correlation matrix is in a recursive form. So, we assume to have

$$\mathbf{R}_{c_x,l}(k) = \alpha_{c_x,l}\mathbf{R}_{c_x,l}(k-1) + (1 - \alpha_{c_x,l})\mathbf{c}_{x,l}(k)\mathbf{c}_{x,l}^T(k). \qquad (9.75)$$

In this case, at time k, $\mathbf{R}_{c_y,l}(k)$ should be computed according to

$$\begin{aligned}\mathbf{R}_{c_y,l}(k) = {}& \alpha_{c_x,l}\mathbf{R}_{c_y,l}(k-1) + (1-\alpha_{c_x,l})\mathbf{c}_{y,l}(k)\mathbf{c}_{y,l}^T(k) + \\ & (\alpha_{c_v,l} - \alpha_{c_x,l})\mathbf{R}_{c_v,l}(k-1) - \\ & (\alpha_{c_v,l} - \alpha_{c_x,l})\mathbf{c}_{v,l}(k)\mathbf{c}_{v,l}^T(k).\end{aligned} \qquad (9.76)$$

9.3 Experiments

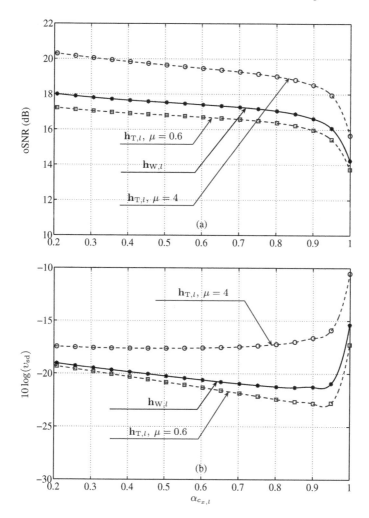

Fig. 9.6. Performance as a function of the forgetting factor $\alpha_{c_x,l}$ in white Gaussian noise (WGN).

The forgetting factors $\alpha_{c_x,l}$ and $\alpha_{c_v,l}$ ($l = 1, \ldots, L$) play an important role in noise reduction performance of the Class-II filters. In principle, each subband l may take a different forgetting factor. But for simplicity, in this study, we assume the same forgetting factor for all the subbands, i.e., $\alpha_{c_x,1} = \alpha_{c_x,2} = \cdots = \alpha_{c_x,L} = \alpha_{c_x}$, and $\alpha_{c_v,1} = \alpha_{c_v,2} = \cdots = \alpha_{c_v,L} = \alpha_{c_v}$. Again, white Gaussian noise is used. We set L_l to 3 and α_{c_v} to 0.91 (this value of α_{c_v} was obtained trough experiments) and examine the noise reduction performance for different values of α_{c_x}. The result of this experiment is plotted in Fig. 9.6. It is seen that, for all the three configured algorithms, the output

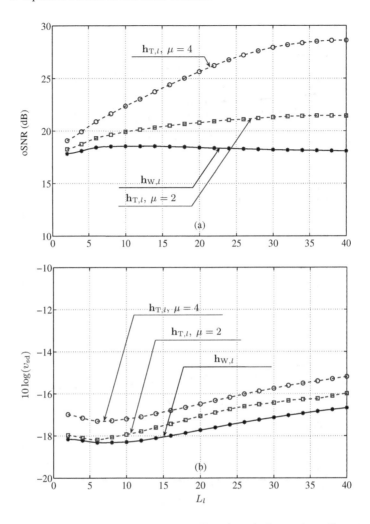

Fig. 9.7. Performance as a function of the filter length L_l in white Gaussian noise (WGN) with an iSNR = 10 dB, $\alpha_{a_y} = 0.8$, $\alpha_{a_v} = 0.91$, and $L = 20$.

SNR monotonically decreases (slowly when $\alpha_{c_x} < 0.9$ and then rapidly) as α_{c_x} increases. In comparison, the speech-distortion index first decreases as α_{c_x} increases and then increases. Taking into account both SNR improvement and speech distortion, one can see that good noise reduction performance is obtained with α_{c_x} being in the range between 0.7 and 0.9.

9.3.6 Effect of Filter Length on Performance of Class-II Filters

In the next experiment, we study the impact of the filter length L_l on the noise reduction performance. Again, the background noise is white and no noise estimator is used. The parameters used in this experiment are $\alpha_{c_x} = 0.8$, $\alpha_{c_v} = 0.91$, iSNR = 10 dB, and $L = 20$. The results are depicted in Fig. 9.7. It is seen from Fig. 9.7(a) that the output SNRs for the tradeoff filters with $\mu = 4$ and $\mu = 2$ increase with L_l. For the Wiener filter, however, the output SNR increases first to its maximum and then decreases. In comparison, the speech-distortion indices with both methods decrease first and then increase with L_l. Taking into account both SNR improvement and speech distortion, we would suggest to use L_l between 5 and 10.

Comparing Figs. 9.7 and 9.2, one can see that, with the same L, the optimal filters in Class II can achieve much higher SNR gain than the filters of Class I. The Class-II filters also have slightly more speech distortion. But the additional amount of distortion compared to that of the Class-I filters is not significant. This indicates that the Class-II filters may have a great potential in practice.

9.4 Summary

In this chapter, we discussed two broad classes of optimal noise reduction filters in the KLE domain. While the filters of Class I achieve a frame of speech estimate by filtering only the corresponding frame of the noisy speech, the filters of Class II are inter-frame techniques, which obtain noise reduction by filtering not only the current frame, but also a number of previous consecutive frames of the noisy speech. We have also discussed some properties and implementation issues of these filters. Through experiments, we have investigated the optimal values of the forgetting factors and the length of the optimal filters. We also demonstrated that better noise reduction performance can be achieved with the filters of Class II when the parameters associated with this class are properly chosen, which demonstrated the great potential of the filters in this category for noise reduction.

It should be pointed out that the underlying idea of the filters of Class II can also be used in the frequency domain, which will be left to the reader's investigation.

10

Optimal Filters in the Transform Domain

In this chapter, we reformulate the noise reduction problem into a more generalized transform domain. There are at least two advantages doing this: first, different transforms can be used to replace each other without any requirement to change the algorithm formulation (optimal or suboptimal filter) and second, it is easier to fairly compare different transforms for their noise reduction performance. To do so, we need to generalize the KLE concept. Therefore, we recommend the reader to be familiar with the KLE (explained in Chapter 2) before reading this part.

10.1 Generalization of the KLE

In this section, we are going to generalize the principle of the KLE to any given unitary transform $\mathbf{U}$. For that, we need to use some of the concepts presented in [12], [106], [107], [129]. The basic idea behind this generalization is to find other ways to exactly diagonalize the correlation matrix $\mathbf{R}_y$[1]. The Fourier matrix, for example, diagonalizes approximately $\mathbf{R}_y$ (since this matrix is Toeplitz and its elements are usually absolutely summable [60]). However, this approximation may cause more distortion to the clean speech when noise reduction is performed in the frequency domain.

We define the square root of the positive definite matrix $\mathbf{R}_y$ as

$$\mathbf{R}_y^{1/2} = \mathbf{Q}_y \mathbf{\Lambda}_y^{1/2} \mathbf{Q}_y^T, \tag{10.1}$$

where the matrices $\mathbf{Q}_y$ and $\mathbf{\Lambda}_y$ are, respectively, orthogonal and diagonal matrices. Expression (10.1) is very useful in the derivation of a generalized form of the KLE.

---

[1] In Chapters 2 and 9, we diagonalized the correlation matrix $\mathbf{R}_x$. In this part, it is far better to diagonalize $\mathbf{R}_y$ instead of $\mathbf{R}_x$ for numerical reasons ($\mathbf{R}_y$ is much better conditioned than $\mathbf{R}_x$) since all optimal filters depend on the inverse of the correlation matrix that we choose to diagonalize.

J. Benesty et al., *Noise Reduction in Speech Processing*, Springer Topics in Signal Processing 2,
DOI 10.1007/978-3-642-00296-0_10, © Springer-Verlag Berlin Heidelberg 2009

124   10 Optimal Filters in the Transform Domain

Consider the $L \times L$ unitary matrix

$$\mathbf{U} = \begin{bmatrix} \mathbf{u}_1 \ \mathbf{u}_2 \ \cdots \ \mathbf{u}_L \end{bmatrix},$$

where $\mathbf{U}^H \mathbf{U} = \mathbf{U} \mathbf{U}^H = \mathbf{I}$ and superscript H denotes transpose conjugate of a vector or a matrix. We would like to minimize the positive quantity $\mathbf{g}^H(\mathbf{u}_l) \mathbf{R}_y^{1/2} \mathbf{g}(\mathbf{u}_l)$ subject to the constraint

$$\mathbf{g}^H(\mathbf{u}_l) \mathbf{u}_l = \mathbf{u}_l^H \mathbf{g}(\mathbf{u}_l) = 1. \tag{10.2}$$

Under this constraint, the process $y(k)$ is passed through the filter

$$\mathbf{g}(\mathbf{u}_l) = \begin{bmatrix} g_0(\mathbf{u}_l) \ g_1(\mathbf{u}_l) \ \cdots \ g_{L-1}(\mathbf{u}_l) \end{bmatrix}^T$$

with no distortion along $\mathbf{u}_l$ and signals along vectors other than $\mathbf{u}_l$ tend to be attenuated. Mathematically, this is equivalent to minimizing the following cost function:

$$J_S\left[\mathbf{g}(\mathbf{u}_l)\right] = \mathbf{g}^H(\mathbf{u}_l) \mathbf{R}_y^{1/2} \mathbf{g}(\mathbf{u}_l) + \mu_S \left[1 - \mathbf{g}^H(\mathbf{u}_l) \mathbf{u}_l\right], \tag{10.3}$$

where μ_S is a Lagrange multiplier. The minimization of (10.3) leads to the following solution:

$$\mathbf{g}(\mathbf{u}_l) = \frac{\mathbf{R}_y^{-1/2} \mathbf{u}_l}{\mathbf{u}_l^H \mathbf{R}_y^{-1/2} \mathbf{u}_l}. \tag{10.4}$$

We define the spectrum of $y(k)$ along $\mathbf{u}_l$ as

$$\phi_y(\mathbf{u}_l) = \mathbf{g}^H(\mathbf{u}_l) \mathbf{R}_y \mathbf{g}(\mathbf{u}_l). \tag{10.5}$$

Substituting (10.4) into (10.5) gives

$$\phi_y(\mathbf{u}_l) = \frac{1}{\left(\mathbf{u}_l^H \mathbf{R}_y^{-1/2} \mathbf{u}_l\right)^2}. \tag{10.6}$$

Expression (10.6) is a general definition of the spectrum of the signal $y(k)$, which depends on the unitary matrix $\mathbf{U}$. Using (10.4) and (10.6), we get

$$\mathbf{R}_y^{1/2} \mathbf{g}(\mathbf{u}_l) = \phi_y^{1/2}(\mathbf{u}_l) \mathbf{u}_l. \tag{10.7}$$

By taking into account all vectors $\mathbf{u}_l$, $l = 1, 2, \ldots, L$, (10.7) can be written into the following general form

$$\mathbf{R}_y^{1/2} \mathbf{G}(\mathbf{U}) = \mathbf{U} \boldsymbol{\Phi}_y^{1/2}(\mathbf{U}), \tag{10.8}$$

where

$$\mathbf{G}(\mathbf{U}) = \begin{bmatrix} \mathbf{g}(\mathbf{u}_1) \ \mathbf{g}(\mathbf{u}_2) \ \cdots \ \mathbf{g}(\mathbf{u}_L) \end{bmatrix}$$

and

$$\boldsymbol{\Phi}_y(\mathbf{U}) = \mathrm{diag}\begin{bmatrix} \phi_y(\mathbf{u}_1) \ \phi_y(\mathbf{u}_2) \ \cdots \ \phi_y(\mathbf{u}_L) \end{bmatrix}$$

is a diagonal matrix.

10.1 Generalization of the KLE 125

Property 10.1. The correlation matrix $\mathbf{R}_y$ can be diagonalized as follows:

$$\mathbf{G}^H(\mathbf{U})\mathbf{R}_y\mathbf{G}(\mathbf{U}) = \mathbf{\Phi}_y(\mathbf{U}). \tag{10.9}$$

Proof. This form follows immediately from (10.8).

Property 10.1 shows that there are an infinite number of ways to diagonalize the matrix $\mathbf{R}_y$, depending on how we choose the unitary matrix $\mathbf{U}$. Each one of these diagonalizations gives a representation of the spectrum of the signal $y(k)$ in the subspace $\mathbf{U}$. Expression (10.9) is a generalization of the KLT; the only major difference is that $\mathbf{G}(\mathbf{U})$ is not a unitary matrix except for the case where $\mathbf{U} = \mathbf{Q}_y$. For this special case, it's easy to verify that $\mathbf{G}(\mathbf{Q}_y) = \mathbf{Q}_y$ and $\mathbf{\Phi}_y(\mathbf{Q}) = \mathbf{\Lambda}_y$, which is the KLT formulation.

Property 10.2. The vector $\mathbf{y}(k)$ can be written as a combination (expansion) of the vectors of the matrix $\mathbf{U}' = \mathbf{R}_y^{1/2}\mathbf{U}\mathbf{\Phi}_y^{-1/2}(\mathbf{U})$ as follows

$$\begin{aligned}
\mathbf{y}(k) &= \sum_{l=1}^{L} c_y\,(k, \mathbf{u}_l)\, \frac{\mathbf{R}_y^{1/2}}{\phi_y^{1/2}(\mathbf{u}_l)}\mathbf{u}_l \\
&= \sum_{l=1}^{L} c_y\,(k, \mathbf{u}_l)\, \mathbf{u}_l',
\end{aligned} \tag{10.10}$$

where

$$c_y\,(k, \mathbf{u}_l) = \mathbf{g}^H(\mathbf{u}_l)\mathbf{y}(k), \ l = 1, 2, \ldots, L \tag{10.11}$$

are the coefficients of the expansion and

$$\mathbf{u}_l' = \frac{\mathbf{R}_y^{1/2}}{\phi_y^{1/2}(\mathbf{u}_l)}\mathbf{u}_l, \ l = 1, 2, \ldots, L. \tag{10.12}$$

Expressions (10.10) and (10.11) are the time- and transform-domain representations of the vector signal $\mathbf{y}(k)$.

Proof. Expressions (10.10) and (10.11) can be shown by substituting one into the other.

Property 10.3. We always have

$$E\,[c_y\,(k, \mathbf{u}_l)] = 0, \ l = 1, 2, \ldots, L \tag{10.13}$$

and

$$E\left[c_y\,(k, \mathbf{u}_i)\, c_y^*\,(k, \mathbf{u}_j)\right] = \begin{cases} \phi_y(\mathbf{u}_i), & i = j \\ 0, & i \neq j \end{cases}. \tag{10.14}$$

Proof. These properties can be verified from (10.11).

126   10 Optimal Filters in the Transform Domain

It can be checked that the Parseval's theorem does not hold anymore if $\mathbf{U} \neq \mathbf{Q}_y$. This is due to the fact that the matrix $\mathbf{G(U)}$ is not unitary. Indeed

$$\sum_{l=1}^{L} E\left[|c_y(k, \mathbf{u}_l)|^2\right] = \sum_{l=1}^{L} \phi_y(\mathbf{u}_l)$$
$$= \mathrm{tr}\left[\mathbf{R}_y \mathbf{G(U)}\mathbf{G}^H(\mathbf{U})\right]$$
$$\neq \mathrm{tr}\left(\mathbf{R}_y\right). \tag{10.15}$$

This is the main difference between the KLT and the generalization proposed here for $\mathbf{U} \neq \mathbf{Q}$. This difference, however, should have no impact on most applications and Properties 10.1, 10.2, and 10.3 are certainly the most important ones.

We define the spectra of the clean speech, $x(k)$, and noise, $v(k)$, in the subspace $\mathbf{U}$ as

$$\phi_x(\mathbf{u}_l) = \mathbf{g}^H(\mathbf{u}_l)\mathbf{R}_x\mathbf{g}(\mathbf{u}_l), \ l = 1, 2, \ldots, L, \tag{10.16}$$
$$\phi_v(\mathbf{u}_l) = \mathbf{g}^H(\mathbf{u}_l)\mathbf{R}_v\mathbf{g}(\mathbf{u}_l), \ l = 1, 2, \ldots, L. \tag{10.17}$$

Of course, $\phi_x(\mathbf{u}_l)$ and $\phi_v(\mathbf{u}_l)$ are always positive real numbers.

We can now apply the three previous properties to our noise reduction problem. Indeed, with the help of Property 10.2 and substituting (2.2) into (10.11), we get

$$c_y(k, \mathbf{u}_l) = \mathbf{g}^H(\mathbf{u}_l)\mathbf{y}(k)$$
$$= \mathbf{g}^H(\mathbf{u}_l)\mathbf{x}(k) + \mathbf{g}^H(\mathbf{u}_l)\mathbf{v}(k)$$
$$= c_x(k, \mathbf{u}_l) + c_v(k, \mathbf{u}_l), \ l = 1, 2, \ldots, L. \tag{10.18}$$

We also have from Property 10.3,

$$E\left[c_y(k, \mathbf{u}_i)c_y^*(k, \mathbf{u}_j)\right] = \begin{cases} \phi_x(\mathbf{u}_i) + \phi_v(\mathbf{u}_i), & i = j \\ 0, & i \neq j \end{cases}. \tag{10.19}$$

Expression (10.18) is equivalent to (2.2) but in the transform domain. Similar to the KLE case, our problem becomes one of finding an estimate of $c_x(k, \mathbf{u}_l)$ by multiplying $c_y(k, \mathbf{u}_l)$ with a (complex) scalar h_l, i.e.,

$$c_z(k, \mathbf{u}_l) = h_l c_y(k, \mathbf{u}_l)$$
$$= h_l\left[c_x(k, \mathbf{u}_l) + c_v(k, \mathbf{u}_l)\right]$$
$$= c_{x,\mathrm{F}}(k, \mathbf{u}_l) + c_{v,\mathrm{F}}(k, \mathbf{u}_l), \ l = 1, 2, \ldots, L. \tag{10.20}$$

The variance of the signal $c_z(k, \mathbf{u}_l)$ is

$$E\left[|c_z(k, \mathbf{u}_l)|^2\right] = |h_l|^2\phi_y(\mathbf{u}_l), \ l = 1, 2, \ldots, L. \tag{10.21}$$

Finally by using Property 10.2, we can go back into the time domain and see that an estimate of the vector $\mathbf{x}(k)$ would be

$$\mathbf{z}(k) = \sum_{l=1}^{L} c_z\left(k, \mathbf{u}_l\right) \mathbf{u}_l'$$

$$= \mathbf{R}_y^{1/2} \left(\sum_{l=1}^{L} h_l \mathbf{u}_l \mathbf{u}_l^H \right) \mathbf{R}_y^{-1/2} \mathbf{y}(k)$$

$$= \mathbf{H}(\mathbf{U}) \left[\mathbf{x}(k) + \mathbf{v}(k) \right], \tag{10.22}$$

where

$$\mathbf{H}(\mathbf{U}) = \mathbf{R}_y^{1/2} \mathbf{U} \ \mathrm{diag} \left[h_1 \ h_2 \ \cdots \ h_L \right] \mathbf{U}^H \mathbf{R}_y^{-1/2} \tag{10.23}$$

is an $L \times L$ (time-domain) filtering matrix, which depends on the unitary matrix $\mathbf{U}$ and is equivalent to the transform-domain filter $\mathbf{h} = \left[h_1 \ h_2 \ \cdots \ h_L \right]^T$. Moreover, it can be checked, with the help of Property 10.1, that the correlation matrix $\mathbf{R}_z = E\left[\mathbf{z}(k) \mathbf{z}^H(k) \right]$ can be diagonalized as follows:

$$\mathbf{G}^H(\mathbf{U}) \mathbf{R}_z \mathbf{G}(\mathbf{U}) = \mathrm{diag} \left[|h_1|^2 \phi_y(\mathbf{u}_1) \ |h_2|^2 \phi_y(\mathbf{u}_2) \ \cdots \ |h_L|^2 \phi_y(\mathbf{u}_L) \right]. \tag{10.24}$$

We see from the previous expression how the coefficients h_l, $l = 1, 2, \ldots, L$, affect the spectrum of the estimated signal $z(k)$ in the subspace $\mathbf{U}$, depending on how they are optimized.

10.2 Performance Measures

Like we did it in the time, frequency, and KLE domains (see Chapter 3), we are going to define the most important measures for noise reduction but in this generalized transform-domain context.

10.2.1 SNR

With the proposed transform-domain representation, we define the subband and fullband input SNRs, respectively, as

$$\mathrm{iSNR}(\mathbf{u}_l) = \frac{E\left[|c_x\left(k, \mathbf{u}_l\right)|^2 \right]}{E\left[|c_v\left(k, \mathbf{u}_l\right)|^2 \right]}$$

$$= \frac{\phi_x(\mathbf{u}_l)}{\phi_v(\mathbf{u}_l)}, \ l = 1, 2, \ldots, L, \tag{10.25}$$

$$\mathrm{iSNR}(\mathbf{U}) = \frac{\sum_{l=1}^{L} \phi_x(\mathbf{u}_l)}{\sum_{l=1}^{L} \phi_v(\mathbf{u}_l)}. \tag{10.26}$$

In general, $\mathrm{iSNR}(\mathbf{U}) \neq \mathrm{iSNR}$. But for $\mathbf{U} = \mathbf{Q}_y$, $\mathrm{iSNR}(\mathbf{Q}_y) = \mathrm{iSNR}$.

After noise reduction with the model given in (10.22), the output SNR is

128   10 Optimal Filters in the Transform Domain

$$\text{oSNR}\left[\mathbf{H}(\mathbf{U})\right] = \frac{\text{tr}\left[\mathbf{H}(\mathbf{U})\mathbf{R}_x\mathbf{H}^H(\mathbf{U})\right]}{\text{tr}\left[\mathbf{H}(\mathbf{U})\mathbf{R}_v\mathbf{H}^H(\mathbf{U})\right]}. \tag{10.27}$$

With the transform-domain model shown in (10.20) and after noise reduction, the subband output SNR is

$$\begin{aligned}\text{oSNR}(\mathbf{u}_l) &= \frac{|h_l|^2\phi_x(\mathbf{u}_l)}{|h_l|^2\phi_v(\mathbf{u}_l)} \\ &= \text{iSNR}(\mathbf{u}_l), \ l = 1, 2, \ldots, L \end{aligned} \tag{10.28}$$

and the fullband output SNR is

$$\text{oSNR}(\mathbf{h}, \mathbf{U}) = \frac{\sum_{l=1}^{L}|h_l|^2\phi_x(\mathbf{u}_l)}{\sum_{l=1}^{L}|h_l|^2\phi_v(\mathbf{u}_l)}. \tag{10.29}$$

In general, $\text{oSNR}(\mathbf{h}, \mathbf{U}) \neq \text{oSNR}\left[\mathbf{H}(\mathbf{U})\right]$. But in the special case where $\mathbf{U} = \mathbf{Q}_y$, we have $\text{oSNR}(\mathbf{h}, \mathbf{Q}_y) = \text{oSNR}\left[\mathbf{H}(\mathbf{Q}_y)\right]$. It is important to observe that the subband output SNR is equal to the subband input SNR; therefore these SNRs are not influenced by the coefficients h_l, $l = 1, 2, \ldots, L$.

Property 10.4. We always have

$$\sum_{l=1}^{L}\text{iSNR}(\mathbf{u}_l) \geq \text{iSNR}(\mathbf{U}), \tag{10.30}$$

$$\sum_{l=1}^{L}\text{oSNR}(\mathbf{u}_l) \geq \text{oSNR}(\mathbf{h}, \mathbf{U}). \tag{10.31}$$

This means that the aggregation of the subband (input or output) SNRs is greater than or equal to the fullband (input or output) SNR.

Proof. The two previous inequalities can be easily shown by using the inequality (3.16).

10.2.2 Noise-Reduction Factor

We define the noise-reduction factor for the model in (10.22) as

$$\xi_{\text{nr}}\left[\mathbf{H}(\mathbf{U})\right] = \frac{\text{tr}\left(\mathbf{R}_v\right)}{\text{tr}\left[\mathbf{H}(\mathbf{U})\mathbf{R}_v\mathbf{H}^H(\mathbf{U})\right]}. \tag{10.32}$$

The larger the value of $\xi_{\text{nr}}\left[\mathbf{H}(\mathbf{U})\right]$, the more the noise is reduced. After the filtering operation, the residual noise level is expected to be lower than that of the original noise level, therefore this factor should be lower bounded by 1. The subband noise-reduction factor is

$$\xi_{\mathrm{nr}}(h_l) = \frac{1}{|h_l|^2}, \ l = 1, 2, \ldots, L \tag{10.33}$$

and the corresponding fullband noise-reduction factor is

$$\xi_{\mathrm{nr}}(\mathbf{h}, \mathbf{U}) = \frac{\sum_{l=1}^{L} \phi_v(\mathbf{u}_l)}{\sum_{l=1}^{L} |h_l|^2 \phi_v(\mathbf{u}_l)}. \tag{10.34}$$

In general, $\xi_{\mathrm{nr}}(\mathbf{h}, \mathbf{U}) \neq \xi_{\mathrm{nr}}[\mathbf{H}(\mathbf{U})]$. But for $\mathbf{U} = \mathbf{Q}_y$, $\xi_{\mathrm{nr}}(\mathbf{h}, \mathbf{Q}_y) = \xi_{\mathrm{nr}}[\mathbf{H}(\mathbf{Q}_y)]$.

Property 10.5. We always have

$$\sum_{l=1}^{L} \xi_{\mathrm{nr}}(h_l) \geq \xi_{\mathrm{nr}}(\mathbf{h}, \mathbf{U}). \tag{10.35}$$

This means that the aggregation of the subband noise-reduction factors is greater than or equal to the fullband noise-reduction factor.

Proof. Easy to show by using the inequality (3.16).

10.2.3 Speech-Distortion Index

The filtering operation adds distortion to the speech signal; so a measure is needed to quantify the amount of speech distortion. With the model given in (10.22), we define the speech-distortion index as

$$\upsilon_{\mathrm{sd}}[\mathbf{H}(\mathbf{U})] = \frac{E\left\{ [\mathbf{x}(k) - \mathbf{H}(\mathbf{U})\mathbf{x}(k)]^H [\mathbf{x}(k) - \mathbf{H}(\mathbf{U})\mathbf{x}(k)] \right\}}{\mathrm{tr}(\mathbf{R}_x)}. \tag{10.36}$$

This index is lower bounded by 0 and expected to be upper bounded by 1 for optimal filters. The higher the value of $\upsilon_{\mathrm{sd}}[\mathbf{H}(\mathbf{U})]$, the more the speech is distorted.

In the transform domain with the formulation given in (10.20), we define the subband and fullband speech-distortion indices, respectively, as

$$\upsilon_{\mathrm{sd}}(h_l) = |1 - h_l|^2, \ l = 1, 2, \ldots, L \tag{10.37}$$

and

$$\upsilon_{\mathrm{sd}}(\mathbf{h}, \mathbf{U}) = \frac{\sum_{l=1}^{L} |1 - h_l|^2 \phi_x(\mathbf{u}_l)}{\sum_{l=1}^{L} \phi_x(\mathbf{u}_l)}. \tag{10.38}$$

In general, $\upsilon_{\mathrm{sd}}(\mathbf{h}, \mathbf{U}) \neq \upsilon_{\mathrm{sd}}[\mathbf{H}(\mathbf{U})]$. But for the special case of $\mathbf{U} = \mathbf{Q}_y$, $\upsilon_{\mathrm{sd}}(\mathbf{h}, \mathbf{Q}_y) = \upsilon_{\mathrm{sd}}[\mathbf{H}(\mathbf{Q}_y)]$.

130   10 Optimal Filters in the Transform Domain

Property 10.6. We always have

$$\sum_{l=1}^{L} \upsilon_{sd}(h_l) \geq \upsilon_{sd}(\mathbf{h}, \mathbf{U}). \tag{10.39}$$

This means that the aggregation of the subband speech-distortion indices is greater than or equal to the fullband speech-distortion index.

Proof. Easy to show by using the inequality (3.16).

10.2.4 Speech-Reduction Factor

This factor measures the amount of speech reduced by the filtering operation. We define the speech-reduction factor for the model in (10.22) as

$$\xi_{sr}[\mathbf{H}(\mathbf{U})] = \frac{\text{tr}(\mathbf{R}_x)}{\text{tr}\left[\mathbf{H}(\mathbf{U})\mathbf{R}_x\mathbf{H}^H(\mathbf{U})\right]}. \tag{10.40}$$

The larger the value of $\xi_{sr}[\mathbf{H}(\mathbf{U})]$, the more the desired speech is reduced. After the filtering operation, the residual speech level is expected to be lower than that of the original speech level, therefore this factor should be lower bounded by 1.

The subband speech-reduction factor is

$$\xi_{sr}(h_l) = \frac{1}{|h_l|^2}, \ l = 1, 2, \ldots, L \tag{10.41}$$

and the corresponding fullband speech-reduction factor is

$$\xi_{sr}(\mathbf{h}, \mathbf{U}) = \frac{\sum_{l=1}^{L} \phi_x(\mathbf{u}_l)}{\sum_{l=1}^{L} |h_l|^2 \phi_x(\mathbf{u}_l)}. \tag{10.42}$$

In general, $\xi_{sr}(\mathbf{h}, \mathbf{U}) \neq \xi_{sr}[\mathbf{H}(\mathbf{U})]$. But for $\mathbf{U} = \mathbf{Q}_y$, $\xi_{sr}(\mathbf{h}, \mathbf{Q}_y) = \xi_{sr}[\mathbf{H}(\mathbf{Q}_y)]$.

Property 10.7. We always have

$$\sum_{l=1}^{L} \xi_{sr}(h_l) \geq \xi_{sr}(\mathbf{h}, \mathbf{U}). \tag{10.43}$$

This means that the aggregation of the subband speech-reduction factors is greater than or equal to the fullband speech-reduction factor.

Proof. Easy to show by using the inequality (3.16).

Property 10.8. We always have

$$\frac{\text{oSNR}[\mathbf{H}(\mathbf{U})]}{\text{iSNR}} = \frac{\xi_{nr}[\mathbf{H}(\mathbf{U})]}{\xi_{sr}[\mathbf{H}(\mathbf{U})]}, \tag{10.44}$$

$$\frac{\text{oSNR}(\mathbf{h}, \mathbf{U})}{\text{iSNR}(\mathbf{U})} = \frac{\xi_{nr}(\mathbf{h}, \mathbf{U})}{\xi_{sr}(\mathbf{h}, \mathbf{U})}. \tag{10.45}$$

Proof. Easy to see by using the different definitions of the involved measures.

10.3 MSE Criterion

Let us define the transform-domain error signal between the estimated and desired signals in the subband l:

$$\begin{aligned} e\left(k, \mathbf{u}_l\right) &= c_z\left(k, \mathbf{u}_l\right) - c_x\left(k, \mathbf{u}_l\right) \\ &= h_l c_y\left(k, \mathbf{u}_l\right) - c_x\left(k, \mathbf{u}_l\right), \ l = 1, 2, \ldots, L. \end{aligned} \tag{10.46}$$

This error can also be written as the sum of two error signals:

$$e\left(k, \mathbf{u}_l\right) = e_x\left(k, \mathbf{u}_l\right) + e_v\left(k, \mathbf{u}_l\right), \tag{10.47}$$

where

$$e_x\left(k, \mathbf{u}_l\right) = (h_l - 1)c_x\left(k, \mathbf{u}_l\right) \tag{10.48}$$

is the speech distortion due to the complex gain, and

$$e_v\left(k, \mathbf{u}_l\right) = h_l c_v\left(k, \mathbf{u}_l\right) \tag{10.49}$$

represents the residual noise.

From the error signal (10.46), we give the corresponding transform-domain MSE criterion in the subband l:

$$\begin{aligned} J\left(h_l, \mathbf{u}_l\right) &= E\left[\left|e\left(k, \mathbf{u}_l\right)\right|^2\right] \\ &= \phi_x(\mathbf{u}_l) + \left|h_l\right|^2 \phi_y(\mathbf{u}_l) - 2\mathcal{R}\left[h_l \phi_{yx}(\mathbf{u}_l)\right] \\ &= \phi_x(\mathbf{u}_l) + \left|h_l\right|^2 \phi_y(\mathbf{u}_l) - 2\mathcal{R}\left[h_l \phi_x(\mathbf{u}_l)\right], \end{aligned} \tag{10.50}$$

where

$$\begin{aligned} \phi_{yx}(\mathbf{u}_l) &= E\left[c_y\left(k, \mathbf{u}_l\right) c_x^*\left(k, \mathbf{u}_l\right)\right] \\ &= \phi_x(\mathbf{u}_l) \end{aligned}$$

is the cross-spectrum between the observation and speech signals. Expression (10.50) can be structured in a different way:

$$\begin{aligned} J\left(h_l, \mathbf{u}_l\right) &= E\left[\left|e_x\left(k, \mathbf{u}_l\right)\right|^2\right] + E\left[\left|e_v\left(k, \mathbf{u}_l\right)\right|^2\right] \\ &= J_x\left(h_l, \mathbf{u}_l\right) + J_v\left(h_l, \mathbf{u}_l\right). \end{aligned} \tag{10.51}$$

For the particular gain $h_l = 1$, $\forall l$, we get

$$J\left(1, \mathbf{u}_l\right) = \phi_v(\mathbf{u}_l), \tag{10.52}$$

so there will be neither noise reduction nor speech distortion. Therefore, we define the transform-domain NMSE in the subband l as

132   10 Optimal Filters in the Transform Domain

$$\tilde{J}(h_l, \mathbf{u}_l) = \frac{J(h_l, \mathbf{u}_l)}{J(1, \mathbf{u}_l)}$$

$$= \mathrm{iSNR}(\mathbf{u}_l) \cdot \upsilon_{\mathrm{sd}}(h_l) + \frac{1}{\xi_{\mathrm{nr}}(h_l)}, \tag{10.53}$$

where

$$\upsilon_{\mathrm{sd}}(h_l) = \frac{J_x(h_l, \mathbf{u}_l)}{\phi_x(\mathbf{u}_l)}, \tag{10.54}$$

$$\xi_{\mathrm{nr}}(h_l) = \frac{\phi_v(\mathbf{u}_l)}{J_v(h_l, \mathbf{u}_l)}. \tag{10.55}$$

The transform-domain NMSE depends explicitly on the subband input SNR, the subband speech-distortion index, and the subband noise-reduction factor.

We define the fullband MSE and fullband NMSE as

$$J(\mathbf{h}, \mathbf{U}) = \frac{1}{L} \sum_{l=1}^{L} J(h_l, \mathbf{u}_l) \tag{10.56}$$

$$= \frac{1}{L} \sum_{l=1}^{L} |h_l - 1|^2 \phi_x(\mathbf{u}_l) + \frac{1}{L} \sum_{l=1}^{L} |h_l|^2 \phi_v(\mathbf{u}_l)$$

$$= J_x(\mathbf{h}, \mathbf{U}) + J_v(\mathbf{h}, \mathbf{U})$$

and

$$\tilde{J}(\mathbf{h}, \mathbf{U}) = L \frac{J(\mathbf{h}, \mathbf{U})}{\sum_{l=1}^{L} \phi_v(\mathbf{u}_l)} \tag{10.57}$$

$$= \frac{\sum_{l=1}^{L} |h_l - 1|^2 \phi_x(\mathbf{u}_l)}{\sum_{l=1}^{L} \phi_v(\mathbf{u}_l)} + \frac{\sum_{l=1}^{L} |h_l|^2 \phi_v(\mathbf{u}_l)}{\sum_{l=1}^{L} \phi_v(\mathbf{u}_l)}$$

$$= \mathrm{iSNR}(\mathbf{U}) \cdot \upsilon_{\mathrm{sd}}(\mathbf{h}, \mathbf{U}) + \frac{1}{\xi_{\mathrm{nr}}(\mathbf{h}, \mathbf{U})},$$

where

$$\upsilon_{\mathrm{sd}}(\mathbf{h}, \mathbf{U}) = \frac{J_x(\mathbf{h}, \mathbf{U})}{\sum_{l=1}^{L} \phi_x(\mathbf{u}_l)}, \tag{10.58}$$

$$\xi_{\mathrm{nr}}(\mathbf{h}, \mathbf{U}) = \frac{\sum_{l=1}^{L} \phi_v(\mathbf{u}_l)}{J_v(\mathbf{h}, \mathbf{U})}. \tag{10.59}$$

The fullband NMSE depends explicitly on the fullband input SNR, the fullband speech-distortion index, and the fullband noise-reduction factor.

10.4 PCC and Fundamental Properties

Let $c_a(k, \mathbf{u}_l)$ and $c_b(k, \mathbf{u}_l)$ be two signals in the transform domain. We define the subband SPCC or the MSCF between $c_a(k, \mathbf{u}_l)$ and $c_b(k, \mathbf{u}_l)$ in the subband l as

$$|\rho\left[c_a\left(k, \mathbf{u}_l\right), c_b\left(k, \mathbf{u}_l\right)\right]|^2 = \frac{\left|E\left[c_a\left(k, \mathbf{u}_l\right) c_b^*\left(k, \mathbf{u}_l\right)\right]\right|^2}{E\left[\left|c_a\left(k, \mathbf{u}_l\right)\right|^2\right] E\left[\left|c_b\left(k, \mathbf{u}_l\right)\right|^2\right]}$$

$$= \frac{|\phi_{ab}(\mathbf{u}_l)|^2}{\phi_a(\mathbf{u}_l)\phi_b(\mathbf{u}_l)}. \tag{10.60}$$

Now, let us define the two vectors of length L:

$$\mathbf{c}_a(k, \mathbf{U}) = \begin{bmatrix} c_a(k, \mathbf{u}_1) \ c_a(k, \mathbf{u}_2) \ \cdots \ c_a(k, \mathbf{u}_L) \end{bmatrix}^T,$$
$$\mathbf{c}_b(k, \mathbf{U}) = \begin{bmatrix} c_b(k, \mathbf{u}_1) \ c_b(k, \mathbf{u}_2) \ \cdots \ c_b(k, \mathbf{u}_L) \end{bmatrix}^T.$$

We define the fullband SPCC as

$$|\rho\left[\mathbf{c}_a\left(k, \mathbf{U}\right), \mathbf{c}_b\left(k, \mathbf{U}\right)\right]|^2 = \frac{\left|E\left[\mathbf{c}_a^T\left(k, \mathbf{U}\right) \mathbf{c}_b^*\left(k, \mathbf{U}\right)\right]\right|^2}{E\left[\mathbf{c}_a^T\left(k, \mathbf{U}\right) \mathbf{c}_a^*\left(k, \mathbf{U}\right)\right] E\left[\mathbf{c}_a^T\left(k, \mathbf{U}\right) \mathbf{c}_a^*\left(k, \mathbf{U}\right)\right]}$$

$$= \frac{\left|\sum_{l=1}^L \phi_{ab}(\mathbf{u}_l)\right|^2}{\sum_{l=1}^L \phi_a(\mathbf{u}_l) \cdot \sum_{l=1}^L \phi_b(\mathbf{u}_l)}. \tag{10.61}$$

It is clear that

$$0 \leq |\rho\left[c_a\left(k, \mathbf{u}_l\right), c_b\left(k, \mathbf{u}_l\right)\right]|^2 \leq 1, \tag{10.62}$$

$$0 \leq |\rho\left[\mathbf{c}_a\left(k, \mathbf{U}\right), \mathbf{c}_b\left(k, \mathbf{U}\right)\right]|^2 \leq 1. \tag{10.63}$$

The MSCF between the two signals $c_x\left(k, \mathbf{u}_l\right)$ and $c_y\left(k, \mathbf{u}_l\right)$ in the subband l is

$$|\rho\left[c_x\left(k, \mathbf{u}_l\right), c_y\left(k, \mathbf{u}_l\right)\right]|^2 = \frac{\phi_x(\mathbf{u}_l)}{\phi_x(\mathbf{u}_l) + \phi_v(\mathbf{u}_l)}$$

$$= \frac{\mathrm{iSNR}(\mathbf{u}_l)}{1 + \mathrm{iSNR}(\mathbf{u}_l)}. \tag{10.64}$$

This MSCF tells us how much the observation signal is noisy in the subband l. A value of $|\rho\left[c_x\left(k, \mathbf{u}_l\right), c_y\left(k, \mathbf{u}_l\right)\right]|^2$ close to 1 implies that the speech is largely dominant while a value of $|\rho\left[c_x\left(k, \mathbf{u}_l\right), c_y\left(k, \mathbf{u}_l\right)\right]|^2$ close to 0 implies that the noise is largely dominant in the subband l.

The fullband SPCC between the two signals $\mathbf{c}_x\left(k, \mathbf{U}\right)$ and $\mathbf{c}_y\left(k, \mathbf{U}\right)$ is

$$|\rho\left[\mathbf{c}_x\left(k, \mathbf{U}\right), \mathbf{c}_y\left(k, \mathbf{U}\right)\right]|^2 = \frac{\sum_{l=1}^L \phi_x(\mathbf{u}_l)}{\sum_{l=1}^L \phi_x(\mathbf{u}_l) + \sum_{l=1}^L \phi_v(\mathbf{u}_l)}$$

$$= \frac{\mathrm{iSNR}(\mathbf{U})}{1 + \mathrm{iSNR}(\mathbf{U})}. \tag{10.65}$$

The subband SPCC between the two signals $c_x\left(k, \mathbf{u}_l\right)$ and $c_z\left(k, \mathbf{u}_l\right)$ in the subband l is

$$|\rho\left[c_x\left(k, \mathbf{u}_l\right), c_z\left(k, \mathbf{u}_l\right)\right]|^2 = |\rho\left[c_x\left(k, \mathbf{u}_l\right), c_y\left(k, \mathbf{u}_l\right)\right]|^2. \tag{10.66}$$

10 Optimal Filters in the Transform Domain

The subband SPCC between $c_x(k, \mathbf{u}_l)$ and $c_z(k, \mathbf{u}_l)$ is equal to the subband SPCC between $c_x(k, \mathbf{u}_l)$ and $c_y(k, \mathbf{u}_l)$ and does not depend on h_l; the same way the subband input SNR is equal to the subband output SNR and does not depend on h_l.

The fullband SPCC between the two signals $\mathbf{c}_x(k, \mathbf{U})$ and $\mathbf{c}_z(k, \mathbf{U})$ is

$$
|\rho\left[\mathbf{c}_x(k, \mathbf{U}), \mathbf{c}_z(k, \mathbf{U})\right]|^2 = \frac{\left|\sum_{l=1}^{L} h_l \phi_x(\mathbf{u}_l)\right|^2}{\left[\sum_{l=1}^{L} \phi_x(\mathbf{u}_l)\right]\left[\sum_{l=1}^{L} |h_l|^2 \phi_y(\mathbf{u}_l)\right]} \quad (10.67)
$$

$$
= \frac{\left|\sum_{l=1}^{L} h_l \phi_x(\mathbf{u}_l)\right|^2}{\left[\sum_{l=1}^{L} \phi_x(\mathbf{u}_l)\right]\left[\sum_{l=1}^{L} |h_l|^2 \phi_x(\mathbf{u}_l)\right]} \times
$$

$$
\frac{\mathrm{oSNR}(\mathbf{h}, \mathbf{U})}{1 + \mathrm{oSNR}(\mathbf{h}, \mathbf{U})}.
$$

Property 10.9. We have

$$
|\rho\left[\mathbf{c}_x(k, \mathbf{U}), \mathbf{c}_z(k, \mathbf{U})\right]|^2 = |\rho\left[\mathbf{c}_x(k, \mathbf{U}), \mathbf{c}_{x,\mathrm{F}}(k, \mathbf{U})\right]|^2 \times
$$

$$
|\rho\left[\mathbf{c}_{x,\mathrm{F}}(k, \mathbf{U}), \mathbf{c}_z(k, \mathbf{U})\right]|^2, \quad (10.68)
$$

where

$$
|\rho\left[\mathbf{c}_x(k, \mathbf{U}), \mathbf{c}_{x,\mathrm{F}}(k, \mathbf{U})\right]|^2 = \frac{\left|\sum_{l=1}^{L} h_l \phi_x(\mathbf{u}_l)\right|^2}{\left[\sum_{l=1}^{L} \phi_x(\mathbf{u}_l)\right]\left[\sum_{l=1}^{L} |h_l|^2 \phi_x(\mathbf{u}_l)\right]} \quad (10.69)
$$

and

$$
|\rho\left[\mathbf{c}_{x,\mathrm{F}}(k, \mathbf{U}), \mathbf{c}_z(k, \mathbf{U})\right]|^2 = \frac{\mathrm{oSNR}(\mathbf{h}, \mathbf{U})}{1 + \mathrm{oSNR}(\mathbf{h}, \mathbf{U})}. \quad (10.70)
$$

The fullband SPCC $|\rho\left[\mathbf{c}_x(k, \mathbf{U}), \mathbf{c}_{x,\mathrm{F}}(k, \mathbf{U})\right]|^2$ is a speech-distortion index. If $h_l = 1$, $\forall l$ (no speech distortion) then $|\rho\left[\mathbf{c}_x(k, \mathbf{U}), \mathbf{c}_{x,\mathrm{F}}(k, \mathbf{U})\right]|^2 = 1$. The closer the value of $|\rho\left[\mathbf{c}_x(k, \mathbf{U}), \mathbf{c}_{x,\mathrm{F}}(k, \mathbf{U})\right]|^2$ is to 0, the more the speech signal is distorted. The fullband SPCC $|\rho\left[\mathbf{c}_{x,\mathrm{F}}(k, \mathbf{U}), \mathbf{c}_z(k, \mathbf{U})\right]|^2$ shows the SNR improvement and reaches its maximum when $\mathrm{oSNR}(\mathbf{h}, \mathbf{U})$ is maximized. The minimization of $|\rho\left[\mathbf{c}_x(k, \mathbf{U}), \mathbf{c}_z(k, \mathbf{U})\right]|^{-2}$ leads to an optimal filter.

Property 10.10. We have

$$
|\rho\left[\mathbf{c}_x(k, \mathbf{U}), \mathbf{c}_z(k, \mathbf{U})\right]|^2 \leq \frac{\mathrm{oSNR}(\mathbf{h}, \mathbf{U})}{1 + \mathrm{oSNR}(\mathbf{h}, \mathbf{U})}, \quad (10.71)
$$

with equality when $h_l = 1$, $\forall l$.

Proof. This property follows immediately from (10.68) since $|\rho\left[\mathbf{c}_x(k, \mathbf{U}), \mathbf{c}_{x,\mathrm{F}}(k, \mathbf{U})\right]|^2 \leq 1$.

10.4 PCC and Fundamental Properties   135

Property 10.11. We have

$$\left|\rho\left[\mathbf{c}_{x,\mathrm{F}}\left(k,\mathbf{U}\right),\mathbf{c}_y\left(k,\mathbf{U}\right)\right]\right|^2 = \left|\rho\left[\mathbf{c}_x\left(k,\mathbf{U}\right),\mathbf{c}_{x,\mathrm{F}}\left(k,\mathbf{U}\right)\right]\right|^2 \times$$
$$\left|\rho\left[\mathbf{c}_x\left(k,\mathbf{U}\right),\mathbf{c}_y\left(k,\mathbf{U}\right)\right]\right|^2. \quad (10.72)$$

Proof. Indeed

$$\left|\rho\left[\mathbf{c}_{x,\mathrm{F}}\left(k,\mathbf{U}\right),\mathbf{c}_y\left(k,\mathbf{U}\right)\right]\right|^2 = \frac{\left|\sum_{l=1}^{L} h_l \phi_x(\mathbf{u}_l)\right|^2}{\left[\sum_{l=1}^{L} |h_l|^2 \phi_x(\mathbf{u}_l)\right]\left[\sum_{l=1}^{L} \phi_y(\mathbf{u}_l)\right]}$$

$$= \frac{\left|\sum_{l=1}^{L} h_l \phi_x(\mathbf{u}_l)\right|^2}{\left[\sum_{l=1}^{L} |h_l|^2 \phi_x(\mathbf{u}_l)\right]\left[\sum_{l=1}^{L} \phi_x(\mathbf{u}_l)\right]} \times$$

$$\frac{\mathrm{iSNR}(\mathbf{U})}{1 + \mathrm{iSNR}(\mathbf{U})}$$

$$= \left|\rho\left[\mathbf{c}_x\left(k,\mathbf{U}\right),\mathbf{c}_{x,\mathrm{F}}\left(k,\mathbf{U}\right)\right]\right|^2 \times$$
$$\left|\rho\left[\mathbf{c}_x\left(k,\mathbf{U}\right),\mathbf{c}_y\left(k,\mathbf{U}\right)\right]\right|^2.$$

Property 10.12. We have

$$\left|\rho\left[\mathbf{c}_{x,\mathrm{F}}\left(k,\mathbf{U}\right),\mathbf{c}_y\left(k,\mathbf{U}\right)\right]\right|^2 \leq \frac{\mathrm{iSNR}(\mathbf{U})}{1 + \mathrm{iSNR}(\mathbf{U})}, \quad (10.73)$$

with equality when $h_l = 1$, $\forall l$.

Proof. This property follows immediately from (10.72) since $\left|\rho\left[\mathbf{c}_x\left(k,\mathbf{U}\right),\mathbf{c}_{x,\mathrm{F}}\left(k,\mathbf{U}\right)\right]\right|^2 \leq 1$.

The MSCF between $c_v\left(k,\mathbf{u}_l\right)$ and $c_y\left(k,\mathbf{u}_l\right)$ is another way to see how much the observation signal is affected by the noise in the subband l. This MSCF is

$$\left|\rho\left[c_v\left(k,\mathbf{u}_l\right),c_y\left(k,\mathbf{u}_l\right)\right]\right|^2 = \frac{\phi_v(\mathbf{u}_l)}{\phi_y(\mathbf{u}_l)}$$

$$= \frac{1}{1 + \mathrm{iSNR}(\mathbf{u}_l)}. \quad (10.74)$$

Property 10.13. We have

$$\left|\rho\left[c_x\left(k,\mathbf{u}_l\right),c_y\left(k,\mathbf{u}_l\right)\right]\right|^2 + \left|\rho\left[c_v\left(k,\mathbf{u}_l\right),c_y\left(k,\mathbf{u}_l\right)\right]\right|^2 = 1 \quad (10.75)$$

and

$$\mathrm{iSNR}(\mathbf{u}_l) = \frac{\left|\rho\left[c_x\left(k,\mathbf{u}_l\right),c_y\left(k,\mathbf{u}_l\right)\right]\right|^2}{\left|\rho\left[c_v\left(k,\mathbf{u}_l\right),c_y\left(k,\mathbf{u}_l\right)\right]\right|^2}. \quad (10.76)$$

136   10 Optimal Filters in the Transform Domain

Proof. Easy to see from (10.64) and (10.74).

The subband input SNR in the subband l is the ratio of two MSCFs and the sum of these two MSCFs never exceeds 1.

The fullband SPCC between $\mathbf{c}_v (k, \mathbf{U})$ and $\mathbf{c}_y (k, \mathbf{U})$ is

$$|\rho \left[\mathbf{c}_v (k, \mathbf{U}) , \mathbf{c}_y (k, \mathbf{U}) \right] |^2 = \frac{1}{1 + \text{iSNR}(\mathbf{U})}. \tag{10.77}$$

The subband SPCC between the two signals $c_v (k, \mathbf{u}_l)$ and $c_z (k, \mathbf{u}_l)$ in the subband l is

$$|\rho \left[c_v (k, \mathbf{u}_l) , c_z (k, \mathbf{u}_l) \right] |^2 = |\rho \left[c_v (k, \mathbf{u}_l) , c_y (k, \mathbf{u}_l) \right] |^2. \tag{10.78}$$

The fullband SPCC between the same signals is

$$\begin{aligned} |\rho \left[\mathbf{c}_v (k, \mathbf{U}) , \mathbf{c}_z (k, \mathbf{U}) \right] |^2 &= \frac{\left| \sum_{l=1}^{L} h_l \phi_v(\mathbf{u}_l) \right|^2}{\left[\sum_{l=1}^{L} \phi_v(\mathbf{u}_l) \right] \left[\sum_{l=1}^{L} |h_l|^2 \phi_y(\mathbf{u}_l) \right]} \tag{10.79} \\ &= \frac{\left| \sum_{l=1}^{L} h_l \phi_v(\mathbf{u}_l) \right|^2}{\left[\sum_{l=1}^{L} \phi_v(\mathbf{u}_l) \right] \left[\sum_{l=1}^{L} |h_l|^2 \phi_v(\mathbf{u}_l) \right]} \times \\ & \quad \frac{1}{1 + \text{oSNR}(\mathbf{h}, \mathbf{U})}. \end{aligned}$$

Property 10.14. We have

$$\begin{aligned} |\rho \left[\mathbf{c}_v (k, \mathbf{U}) , \mathbf{c}_z (k, \mathbf{U}) \right] |^2 = |\rho \left[\mathbf{c}_v (k, \mathbf{U}) , \mathbf{c}_{v,\text{F}} (k, \mathbf{U}) \right] |^2 \times \\ |\rho \left[\mathbf{c}_{v,\text{F}} (k, \mathbf{U}) , \mathbf{c}_z (k, \mathbf{U}) \right] |^2, \quad (10.80) \end{aligned}$$

where

$$|\rho \left[\mathbf{c}_v (k, \mathbf{U}) , \mathbf{c}_{v,\text{F}} (k, \mathbf{U}) \right] |^2 = \frac{\left| \sum_{l=1}^{L} h_l \phi_v(\mathbf{u}_l) \right|^2}{\left[\sum_{l=1}^{L} \phi_v(\mathbf{u}_l) \right] \left[\sum_{l=1}^{L} |h_l|^2 \phi_v(\mathbf{u}_l) \right]} \tag{10.81}$$

and

$$|\rho \left[\mathbf{c}_{v,\text{F}} (k, \mathbf{U}) , \mathbf{c}_z (k, \mathbf{U}) \right] |^2 = \frac{1}{1 + \text{oSNR}(\mathbf{h}, \mathbf{U})}. \tag{10.82}$$

Property 10.15. We have

$$|\rho \left[\mathbf{c}_v (k, \mathbf{U}) , \mathbf{c}_z (k, \mathbf{U}) \right] |^2 \leq \frac{1}{1 + \text{oSNR}(\mathbf{h}, \mathbf{U})}, \tag{10.83}$$

with equality when $h_l = 1$, $\forall l$.

Proof. This property follows immediately from (10.80) since $|\rho\left[\mathbf{c}_v\left(k,\mathbf{U}\right),\mathbf{c}_{v,\mathrm{F}}\left(k,\mathbf{U}\right)\right]|^2 \le 1$.

Property 10.16. We have

$$|\rho\left[\mathbf{c}_{v,\mathrm{F}}\left(k,\mathbf{U}\right),\mathbf{c}_y\left(k,\mathbf{U}\right)\right]|^2 = |\rho\left[\mathbf{c}_v\left(k,\mathbf{U}\right),\mathbf{c}_{v,\mathrm{F}}\left(k,\mathbf{U}\right)\right]|^2 \times$$
$$|\rho\left[\mathbf{c}_v\left(k,\mathbf{U}\right),\mathbf{c}_y\left(k,\mathbf{U}\right)\right]|^2. \quad (10.84)$$

Proof. Indeed

$$|\rho\left[\mathbf{c}_{v,\mathrm{F}}\left(k,\mathbf{U}\right),\mathbf{c}_y\left(k,\mathbf{U}\right)\right]|^2 = \frac{\left|\sum_{l=1}^L h_l\phi_v(\mathbf{u}_l)\right|^2}{\left[\sum_{l=1}^L |h_l|^2 \phi_v(\mathbf{u}_l)\right]\left[\sum_{l=1}^L \phi_y(\mathbf{u}_l)\right]}$$

$$= \frac{\left|\sum_{l=1}^L h_l\phi_v(\mathbf{u}_l)\right|^2}{\left[\sum_{l=1}^L |h_l|^2 \phi_v(\mathbf{u}_l)\right]\left[\sum_{l=1}^L \phi_v(\mathbf{u}_l)\right]} \times$$

$$\frac{1}{1+\mathrm{iSNR}(\mathbf{U})}$$

$$= |\rho\left[\mathbf{c}_v\left(k,\mathbf{U}\right),\mathbf{c}_{v,\mathrm{F}}\left(k,\mathbf{U}\right)\right]|^2 \times$$
$$|\rho\left[\mathbf{c}_v\left(k,\mathbf{U}\right),\mathbf{c}_y\left(k,\mathbf{U}\right)\right]|^2.$$

Property 10.17. We have

$$|\rho\left[\mathbf{c}_{v,\mathrm{F}}\left(k,\mathbf{U}\right),\mathbf{c}_y\left(k,\mathbf{U}\right)\right]|^2 \le \frac{1}{1+\mathrm{iSNR}(\mathbf{U})}, \quad (10.85)$$

with equality when $h_l = 1$, $\forall l$.

Proof. This property follows immediately from (10.84) since $|\rho\left[\mathbf{c}_v\left(k,\mathbf{U}\right),\mathbf{c}_{v,\mathrm{F}}\left(k,\mathbf{U}\right)\right]|^2 \le 1$.

Property 10.18. We have

$$|\rho\left[\mathbf{c}_x\left(k,\mathbf{U}\right),\mathbf{c}_y\left(k,\mathbf{U}\right)\right]|^2 + |\rho\left[\mathbf{c}_v\left(k,\mathbf{U}\right),\mathbf{c}_y\left(k,\mathbf{U}\right)\right]|^2 = 1 \quad (10.86)$$

and

$$\mathrm{iSNR}(\mathbf{U}) = \frac{|\rho\left[\mathbf{c}_x\left(k,\mathbf{U}\right),\mathbf{c}_y\left(k,\mathbf{U}\right)\right]|^2}{|\rho\left[\mathbf{c}_v\left(k,\mathbf{U}\right),\mathbf{c}_y\left(k,\mathbf{U}\right)\right]|^2}. \quad (10.87)$$

Proof. Easy to see from (10.65) and (10.77).

Property 10.19. We have

$$|\rho\left[\mathbf{c}_{x,\mathrm{F}}\left(k,\mathbf{U}\right),\mathbf{c}_z\left(k,\mathbf{U}\right)\right]|^2 + |\rho\left[\mathbf{c}_{v,\mathrm{F}}\left(k,\mathbf{U}\right),\mathbf{c}_z\left(k,\mathbf{U}\right)\right]|^2 = 1 \quad (10.88)$$

and

$$\mathrm{oSNR}(\mathbf{h},\mathbf{U}) = \frac{|\rho\left[\mathbf{c}_{x,\mathrm{F}}\left(k,\mathbf{U}\right),\mathbf{c}_z\left(k,\mathbf{U}\right)\right]|^2}{|\rho\left[\mathbf{c}_{v,\mathrm{F}}\left(k,\mathbf{U}\right),\mathbf{c}_z\left(k,\mathbf{U}\right)\right]|^2}. \quad (10.89)$$

138   10 Optimal Filters in the Transform Domain

Proof. Easy to see from (10.70) and (10.82).

The fullband input or output SNR for the transform-domain approach is the ratio of two fullband SPCCs and the sum of these two SPCCs never exceeds 1.

It can be checked that $\mathrm{oSNR}(\mathbf{h}, \mathbf{U}) > \mathrm{iSNR}(\mathbf{U})$ is equivalent to $|\rho\left[\mathbf{c}_{x,\mathrm{F}}\left(k, \mathbf{U}\right), \mathbf{c}_z\left(k, \mathbf{U}\right)\right]|^2 > \rho^2\left[\mathbf{c}_x\left(k, \mathbf{U}\right), \mathbf{c}_y\left(k, \mathbf{U}\right)\right]$. In this situation, $|\rho\left[\mathbf{c}_{v,\mathrm{F}}\left(k, \mathbf{U}\right), \mathbf{c}_z\left(k, \mathbf{U}\right)\right]|^2 < \rho^2\left[\mathbf{c}_v\left(k, \mathbf{U}\right), \mathbf{c}_y\left(k, \mathbf{U}\right)\right]$.

10.5 Examples of Filter Design

In this section, we develop some important optimal and suboptimal filters in the transform domain.

10.5.1 Wiener Filter

Taking the gradient of $J\left(h_l, \mathbf{u}_l\right)$ with respect to h_l^* and equating the result to 0 lead to

$$-E\left\{c_y^*(k, \mathbf{u}_l)\left[c_x(k, \mathbf{u}_l) - h_{\mathrm{W},l}c_y(k, \mathbf{u}_l)\right]\right\} = 0. \tag{10.90}$$

Hence

$$\phi_y(\mathbf{u}_l)h_{\mathrm{W},l} = \phi_{xy}(\mathbf{u}_l). \tag{10.91}$$

But

$$\phi_{xy}(\mathbf{u}_l) = E\left[c_x(k, \mathbf{u}_l)c_y^*(k, \mathbf{u}_l)\right]$$
$$= \phi_x(\mathbf{u}_l).$$

Therefore the optimal filter can be put into the following forms:

$$h_{\mathrm{W},l} = \frac{\phi_x(\mathbf{u}_l)}{\phi_y(\mathbf{u}_l)}$$
$$= 1 - \frac{\phi_v(\mathbf{u}_l)}{\phi_y(\mathbf{u}_l)}$$
$$= \frac{\mathrm{iSNR}(\mathbf{u}_l)}{1 + \mathrm{iSNR}(\mathbf{u}_l)}. \tag{10.92}$$

We see that the transform-domain Wiener filter is always real and positive and its form is similar to that of the frequency-domain Wiener filter (see Chapter 8).

Another way to define the Wiener filter is with the MSCFs. Indeed, it is easy to verify that

$$h_{\mathrm{W},l} = |\rho\left[c_x(k, \mathbf{u}_l), c_y(k, \mathbf{u}_l)\right]|^2$$
$$= 1 - |\rho\left[c_v(k, \mathbf{u}_l), c_y(k, \mathbf{u}_l)\right]|^2. \tag{10.93}$$

Now, let us define the complex number[2]

---

[2] Notice that both $\rho\left[c_x(k, \mathbf{u}_l), c_y(k, \mathbf{u}_l)\right]$ and $\rho\left[c_v(k, \mathbf{u}_l), c_y(k, \mathbf{u}_l)\right]$ are real numbers.

$$\varrho\left[c_x(k,\mathbf{u}_l),c_v(k,\mathbf{u}_l)\right] = \rho\left[c_x(k,\mathbf{u}_l),c_y(k,\mathbf{u}_l)\right] + j\rho\left[c_v(k,\mathbf{u}_l),c_y(k,\mathbf{u}_l)\right]$$
$$= \cos\theta(\mathbf{u}_l) + j\sin\theta(\mathbf{u}_l), \tag{10.94}$$

where $\theta(\mathbf{u}_l)$ is the angle of $\varrho\left[c_x(k,\mathbf{u}_l),c_v(k,\mathbf{u}_l)\right]$ for which the modulus is equal to 1. On the complex plane, $\varrho\left[c_x(k,\mathbf{u}_l),c_v(k,\mathbf{u}_l)\right]$ is on the unit circle. Since $0 \le \rho\left[c_x(k,\mathbf{u}_l),c_y(k,\mathbf{u}_l)\right] \le 1$ and $0 \le \rho\left[c_v(k,\mathbf{u}_l),c_y(k,\mathbf{u}_l)\right] \le 1$, therefore $0 \le \theta(\mathbf{u}_l) \le \frac{\pi}{2}$. We can then rewrite the Wiener filter as a function of the angle $\theta(\mathbf{u}_l)$:

$$h_{\mathrm{W},l} = \cos^2\theta(\mathbf{u}_l)$$
$$= 1 - \sin^2\theta(\mathbf{u}_l). \tag{10.95}$$

Hence

$$\lim_{\theta(\mathbf{u}_l)\to 0} h_{\mathrm{W},l} = 1, \tag{10.96}$$

$$\lim_{\theta(\mathbf{u}_l)\to\frac{\pi}{2}} h_{\mathrm{W},l} = 0. \tag{10.97}$$

We deduce the subband noise-reduction factor and subband speech-distortion index

$$\xi_{\mathrm{nr}}\left(h_{\mathrm{W},l}\right) = \frac{1}{\cos^4\theta(\mathbf{u}_l)} \ge 1, \tag{10.98}$$

$$\upsilon_{\mathrm{sd}}\left(h_{\mathrm{W},l}\right) = \sin^4\theta(\mathbf{u}_l) \le 1, \tag{10.99}$$

and the subband MNMSE

$$\tilde{J}\left(h_{\mathrm{W},l},\mathbf{u}_l\right) = h_{\mathrm{W},l}, \tag{10.100}$$

which is exactly the transform-domain Wiener filter. We see clearly how noise reduction and speech distortion depend on the angle $\theta(\mathbf{u}_l)$ in the Wiener filter. When $\theta(\mathbf{u}_l)$ increases so does $\xi_{\mathrm{nr}}\left(h_{\mathrm{W},l}\right)$; at the same time $\upsilon_{\mathrm{sd}}\left(h_{\mathrm{W},l}\right)$ increases.

Property 10.20. With the optimal transform-domain Wiener filter given in (10.92), the fullband output SNR is always greater than or equal to the fullband input SNR, i.e., $\mathrm{oSNR}(\mathbf{h}_{\mathrm{W}},\mathbf{U}) \ge \mathrm{iSNR}(\mathbf{U})$.

Proof. Let us evaluate the fullband SPCC between $\mathbf{c}_y(k,\mathbf{U})$ and $\mathbf{c}_{z,\mathrm{W}}(k,\mathbf{U}) = \left[h_{\mathrm{W},1}c_y(k,\mathbf{u}_1)\ h_{\mathrm{W},2}c_y(k,\mathbf{u}_2)\ \cdots\ h_{\mathrm{W},L}c_y(k,\mathbf{u}_L)\right]^T$:

$$\left|\rho\left[\mathbf{c}_y(k,\mathbf{U}),\mathbf{c}_{z,\mathrm{W}}(k,\mathbf{U})\right]\right|^2 = \frac{\left[\sum_{l=1}^{L} h_{\mathrm{W},l}\phi_y(\mathbf{u}_l)\right]^2}{\left[\sum_{l=1}^{L}\phi_y(\mathbf{u}_l)\right]\left[\sum_{l=1}^{L} h_{\mathrm{W},l}^2\phi_y(\mathbf{u}_l)\right]}$$

$$= \frac{\sum_{l=1}^{L}\phi_x(\mathbf{u}_l)}{\sum_{l=1}^{L}\phi_y(\mathbf{u}_l)} \cdot \frac{\sum_{l=1}^{L}\phi_x(\mathbf{u}_l)}{\sum_{l=1}^{L} h_{\mathrm{W},l}\phi_x(\mathbf{u}_l)}$$

$$= \frac{\left|\rho\left[\mathbf{c}_x(k,\mathbf{U}),\mathbf{c}_y(k,\mathbf{U})\right]\right|^2}{\left|\rho\left[\mathbf{c}_y(k,\mathbf{U}),\mathbf{c}_{z,\mathrm{W}}(k,\mathbf{U})\right]\right|^2}.$$

140   10 Optimal Filters in the Transform Domain

Therefore

$$|\rho\left[\mathbf{c}_x(k,\mathbf{U}),\mathbf{c}_y(k,\mathbf{U})\right]|^2 = |\rho\left[\mathbf{c}_y(k,\mathbf{U}),\mathbf{c}_{z,\mathrm{W}}(k,\mathbf{U})\right]|^2 \times$$
$$|\rho\left[\mathbf{c}_x(k,\mathbf{U}),\mathbf{c}_{z,\mathrm{W}}(k,\mathbf{U})\right]|^2$$
$$\leq |\rho\left[\mathbf{c}_x(k,\mathbf{U}),\mathbf{c}_{z,\mathrm{W}}(k,\mathbf{U})\right]|^2.$$

Using (10.65) and Property 10.10 in the previous expression, we get

$$\frac{\mathrm{iSNR}(\mathbf{U})}{1+\mathrm{iSNR}(\mathbf{U})} \leq \frac{\mathrm{oSNR}(\mathbf{h}_\mathrm{W},\mathbf{U})}{1+\mathrm{oSNR}(\mathbf{h}_\mathrm{W},\mathbf{U})},$$

as a result

$$\mathrm{oSNR}(\mathbf{h}_\mathrm{W},\mathbf{U}) \geq \mathrm{iSNR}(\mathbf{U}).$$

Substituting (10.92) into (10.57), we find the fullband MNMSE:

$$\tilde{J}\left(\mathbf{h}_\mathrm{W},\mathbf{U}\right) = 1 - \frac{\sum_{l=1}^{L}\phi_v^2(\mathbf{u}_l)\phi_y^{-1}(\mathbf{u}_l)}{\sum_{l=1}^{L}\phi_v(\mathbf{u}_l)} \leq 1. \tag{10.101}$$

We can compute the fullband speech-distortion index by substituting (10.92) into (10.38):

$$\upsilon_{\mathrm{sd}}\left(\mathbf{h}_\mathrm{W},\mathbf{U}\right) = 1 - \frac{\mathrm{oSNR}(\mathbf{h}_\mathrm{W},\mathbf{U})+2}{\mathrm{iSNR}(\mathbf{U})\cdot\xi_{\mathrm{nr}}\left(\mathbf{h}_\mathrm{W},\mathbf{U}\right)} \leq 1. \tag{10.102}$$

Using (10.57) and (10.102), we get the fullband noise-reduction factor:

$$\xi_{\mathrm{nr}}\left(\mathbf{h}_\mathrm{W},\mathbf{U}\right) = \frac{\mathrm{oSNR}(\mathbf{h}_\mathrm{W},\mathbf{U})+1}{\mathrm{iSNR}(\mathbf{U})-\tilde{J}\left(\mathbf{h}_\mathrm{W},\mathbf{U}\right)} \geq 1. \tag{10.103}$$

Property 10.21. We have

$$\tilde{J}\left(\mathbf{h}_\mathrm{W},\mathbf{U}\right) = \mathrm{iSNR}(\mathbf{U})\left\{1 - |\rho[\mathbf{c}_x(k,\mathbf{U}),\mathbf{c}_{z,\mathrm{W}}(k,\mathbf{U})]|^2\right\}. \tag{10.104}$$

Proof. Indeed

$$\tilde{J}\left(\mathbf{h}_\mathrm{W},\mathbf{U}\right) = \frac{\sum_{l=1}^{L}\phi_x(\mathbf{u}_l)}{\sum_{l=1}^{L}\phi_v(\mathbf{u}_l)} - \frac{\sum_{l=1}^{L}\phi_x^2(\mathbf{u}_l)\phi_y^{-1}(\mathbf{u}_l)}{\sum_{l=1}^{L}\phi_v(\mathbf{u}_l)}$$
$$= \mathrm{iSNR}(\mathbf{U})\left[1 - \frac{\sum_{l=1}^{L}\phi_x^2(\mathbf{u}_l)\phi_y^{-1}(\mathbf{u}_l)}{\sum_{l=1}^{L}\phi_x(\mathbf{u}_l)}\right]$$
$$= \mathrm{iSNR}(\mathbf{U})\left\{1 - |\rho[\mathbf{c}_x(k,\mathbf{U}),\mathbf{c}_{z,\mathrm{W}}(k,\mathbf{U})]|^2\right\}.$$

Therefore, the fullband NMSE is minimized when the fullband SPCC between the signals $\mathbf{c}_x(k,\mathbf{U})$ and $\mathbf{c}_z(k,\mathbf{U})$ is maximized. This fullband SPCC can be rewritten as follows:

$$|\rho[\mathbf{c}_x(k,\mathbf{U}),\mathbf{c}_{z,\mathrm{W}}(k,\mathbf{U})]|^2 = \frac{1}{\xi_{\mathrm{sr}}(\mathbf{h}_\mathrm{W},\mathbf{U})} \cdot \frac{1+\mathrm{oSNR}(\mathbf{h}_\mathrm{W},\mathbf{U})}{\mathrm{oSNR}(\mathbf{h}_\mathrm{W},\mathbf{U})}. \quad (10.105)$$

We observe that the transform-domain Wiener filter is compromising between speech reduction (i.e., speech distortion) and fullband output SNR improvement.

Property 10.22. We have

$$\frac{\mathrm{iSNR}(\mathbf{U})}{1+\mathrm{oSNR}(\mathbf{h}_\mathrm{W},\mathbf{U})} \le \tilde{J}(\mathbf{h}_\mathrm{W},\mathbf{U}) \le \frac{\mathrm{iSNR}(\mathbf{U})}{1+\mathrm{iSNR}(\mathbf{U})}. \quad (10.106)$$

Proof. Since

$$|\rho[\mathbf{c}_x(k,\mathbf{U}),\mathbf{c}_{z,\mathrm{W}}(k,\mathbf{U})]|^2 \ge \frac{\mathrm{iSNR}(\mathbf{U})}{1+\mathrm{iSNR}(\mathbf{U})}$$

and with the help of (10.104), we easily get

$$\tilde{J}(\mathbf{h}_\mathrm{W},\mathbf{U}) \le \frac{\mathrm{iSNR}(\mathbf{U})}{1+\mathrm{iSNR}(\mathbf{U})}.$$

Since

$$|\rho[\mathbf{c}_x(k,\mathbf{U}),\mathbf{c}_{z,\mathrm{W}}(k,\mathbf{U})]|^2 \le \frac{\mathrm{oSNR}(\mathbf{h}_\mathrm{W},\mathbf{U})}{1+\mathrm{oSNR}(\mathbf{h}_\mathrm{W},\mathbf{U})}$$

and, again, with the help of (10.104), we obtain

$$\frac{\mathrm{iSNR}(\mathbf{U})}{1+\mathrm{oSNR}(\mathbf{h}_\mathrm{W},\mathbf{U})} \le \tilde{J}(\mathbf{h}_\mathrm{W},\mathbf{U}).$$

Property 10.23. We have

$$\frac{[1+\mathrm{oSNR}(\mathbf{h}_\mathrm{W},\mathbf{U})]^2}{\mathrm{iSNR}(\mathbf{U})\cdot\mathrm{oSNR}(\mathbf{h}_\mathrm{W},\mathbf{U})} \le \xi_{\mathrm{nr}}(\mathbf{h}_\mathrm{W},\mathbf{U}) \le$$
$$\frac{[1+\mathrm{iSNR}(\mathbf{U})][1+\mathrm{oSNR}(\mathbf{h}_\mathrm{W},\mathbf{U})]}{\mathrm{iSNR}^2(\mathbf{U})}. \quad (10.107)$$

Proof. Easy to show by using (10.103) and the bounds of $\tilde{J}(\mathbf{h}_\mathrm{W},\mathbf{U})$ [eq. (10.106)].

Property 10.24. We have

$$\frac{1}{[1+\mathrm{oSNR}(\mathbf{h}_\mathrm{W},\mathbf{U})]^2} \le \upsilon_{\mathrm{sd}}(\mathbf{h}_\mathrm{W},\mathbf{U}) \le \frac{1+\mathrm{oSNR}(\mathbf{h}_\mathrm{W},\mathbf{U})-\mathrm{iSNR}(\mathbf{U})}{[1+\mathrm{iSNR}(\mathbf{U})][1+\mathrm{oSNR}(\mathbf{h}_\mathrm{W},\mathbf{U})]}.$$
$$(10.108)$$

142   10 Optimal Filters in the Transform Domain

Proof. Easy to show by using (10.102) and the bounds of $\xi_{\mathrm{nr}}(\mathbf{h_W}, \mathbf{U})$ [eq. (10.107)].

To finish this study on the Wiener filter, let us convert the transform-domain Wiener filter, $\mathbf{h_W}$, into the time-domain filtering matrix, $\mathbf{H_W}(\mathbf{U})$. Indeed, with (10.23) and (10.92) we can rewrite, equivalently, the transform-domain Wiener filter into the time domain:

$$\mathbf{H_W}(\mathbf{U}) = \mathbf{R}_y^{1/2}\mathbf{U}\left[\mathbf{I} - \mathbf{\Phi}_y^{-1}(\mathbf{U})\mathbf{\Phi}_v(\mathbf{U})\right]\mathbf{U}^H\mathbf{R}_y^{-1/2}, \qquad (10.109)$$

where

$$\mathbf{\Phi}_v(\mathbf{U}) = \mathrm{diag}\left[\mathbf{G}^H(\mathbf{U})\mathbf{R}_v\mathbf{G}(\mathbf{U})\right] \qquad (10.110)$$

is a diagonal matrix whose nonzero elements are the elements of the diagonal of the matrix $\mathbf{G}^H(\mathbf{U})\mathbf{R}_v\mathbf{G}(\mathbf{U})$. Now if we substitute (10.9) into (7.1) (see Chapter 7), the time-domain Wiener filter [given in (7.1)] can be written as

$$\mathbf{H_W} = \mathbf{R}_y^{1/2}\mathbf{U}\left\{\mathbf{I} - \mathbf{\Phi}_y^{-1/2}(\mathbf{U})\left[\mathbf{G}(\mathbf{U})^H\mathbf{R}_v\mathbf{G}(\mathbf{U})\right]\mathbf{\Phi}_y^{-1/2}(\mathbf{U})\right\}\mathbf{U}^H\mathbf{R}_y^{-1/2}. \qquad (10.111)$$

It is clearly seen that if the matrix $\mathbf{G}^H(\mathbf{U})\mathbf{R}_v\mathbf{G}(\mathbf{U})$ is diagonal, the two filters $\mathbf{H_W}$ and $\mathbf{H_W}(\mathbf{U})$ are identical. In this scenario, it would not matter which unitary matrix we choose.

10.5.2 Parametric Wiener Filter

One convenient way to control the compromise between noise reduction and speech distortion is via the so-called parametric Wiener filter:

$$h_{\mathrm{G},l} = \left[1 - \sin^{\beta_1}\theta(\mathbf{u}_l)\right]^{\beta_2}, \qquad (10.112)$$

where β_1 and β_2 are two positive parameters that allow the control of this compromise. For $(\beta_1, \beta_2) = (2, 1)$, we get the transform-domain Wiener filter developed previously. Taking $(\beta_1, \beta_2) = (2, 1/2)$, leads to

$$h_{\mathrm{P},l} = \sqrt{1 - \sin^2\theta(\mathbf{u}_l)} \qquad (10.113)$$
$$= \cos\theta(\mathbf{u}_l),$$

which is the equivalent form of the power subtraction method. The pair $(\beta_1, \beta_2) = (1, 1)$ gives the equivalent form of the magnitude subtraction method:

$$h_{\mathrm{M},l} = 1 - \sin\theta(\mathbf{u}_l) \qquad (10.114)$$
$$= 1 - \sqrt{1 - \cos^2\theta(\mathbf{u}_l)}.$$

10.5 Examples of Filter Design   143

We can verify that the subband noise-reduction factors for the transform-domain power subtraction and magnitude subtraction methods are

$$\xi_{\mathrm{nr}}\left(h_{\mathrm{P},l}\right) = \frac{1}{\cos^2\theta(\mathbf{u}_l)}, \tag{10.115}$$

$$\xi_{\mathrm{nr}}\left(h_{\mathrm{M},l}\right) = \frac{1}{[1 - \sin\theta(\mathbf{u}_l)]^2}, \tag{10.116}$$

and the corresponding subband speech-distortion indices are

$$\upsilon_{\mathrm{sd}}\left(h_{\mathrm{P},l}\right) = [1 - \cos\theta(\mathbf{u}_l)]^2, \tag{10.117}$$

$$\upsilon_{\mathrm{sd}}\left(h_{\mathrm{M},l}\right) = \sin^2\theta(\mathbf{u}_l). \tag{10.118}$$

We can also easily check that

$$\xi_{\mathrm{nr}}\left(h_{\mathrm{M},l}\right) \geq \xi_{\mathrm{nr}}\left(h_{\mathrm{W},l}\right) \geq \xi_{\mathrm{nr}}\left(h_{\mathrm{P},l}\right), \tag{10.119}$$

$$\upsilon_{\mathrm{sd}}\left(h_{\mathrm{P},l}\right) \leq \upsilon_{\mathrm{sd}}\left(h_{\mathrm{W},l}\right) \leq \upsilon_{\mathrm{sd}}\left(h_{\mathrm{M},l}\right). \tag{10.120}$$

The two previous inequalities are very important from a practical point of view. They show that, among the three methods, the magnitude subtraction is the most aggressive one as far as noise reduction is concerned but at the same time it's the one that will likely distorts most the speech signal. The smoother approach is the power subtraction while the Wiener filter is between the two others in terms of speech distortion and noise reduction. Since $0 \leq h_{\mathrm{G},l} \leq 1$, then $\mathrm{oSNR}\left(\mathbf{h}_{\mathrm{G}}, \mathbf{U}\right) \geq \mathrm{iSNR}(\mathbf{U})$. Therefore, all three methods improve the (fullband) output SNR.

The two particular transform-domain filters derived above can be rewritten, equivalently, into the time domain.

- Power subtraction:

$$\mathbf{H}_{\mathrm{P}}(\mathbf{U}) = \mathbf{R}_y^{1/2}\mathbf{U}\left[\mathbf{I} - \mathbf{\Phi}_y^{-1}(\mathbf{U})\mathbf{\Phi}_v(\mathbf{U})\right]^{1/2}\mathbf{U}^H\mathbf{R}_y^{-1/2}. \tag{10.121}$$

- Magnitude subtraction:

$$\mathbf{H}_{\mathrm{M}}(\mathbf{U}) = \mathbf{R}_y^{1/2}\mathbf{U}\left[\mathbf{I} - \mathbf{\Phi}_y^{-1/2}(\mathbf{U})\mathbf{\Phi}_v^{1/2}(\mathbf{U})\right]\mathbf{U}^H\mathbf{R}_y^{-1/2}. \tag{10.122}$$

These two filters are, of course, not optimal in any sense but they can be very practical.

10.5.3 Tradeoff Filter

Another useful filter can be designed by minimizing the speech distortion with the constraint that the residual noise is equal to a positive value smaller than the level of the original noise. This optimization problem can be translated mathematically as

144   10 Optimal Filters in the Transform Domain

$$\min_{h_l} J_x\left(h_l, \mathbf{u}_l\right) \quad \text{subject to} \quad J_v\left(h_l, \mathbf{u}_l\right) = \beta_l \phi_v(\mathbf{u}_l), \; l = 1, 2, \ldots, L,$$

(10.123)

where $0 < \beta_l < 1$ in order to have some noise reduction. Using a Lagrange multiplier, $\mu_l (\geq 0)$, to adjoin the constraint to the cost function, we can derive the optimal filter:

$$
\begin{aligned}
h_{\mathrm{T},l} &= \frac{\phi_x(\mathbf{u}_l)}{\phi_x(\mathbf{u}_l) + \mu_l \phi_v(\mathbf{u}_l)} \\
&= \frac{\phi_y(\mathbf{u}_l) - \phi_v(\mathbf{u}_l)}{\phi_y(\mathbf{u}_l) + (\mu_l - 1)\phi_v(\mathbf{u}_l)} \\
&= \frac{1 - |\rho\left[c_v\left(k, \mathbf{u}_l\right), c_y\left(k, \mathbf{u}_l\right)\right]|^2}{1 + (\mu_l - 1)|\rho\left[c_v\left(k, \mathbf{u}_l\right), c_y\left(k, \mathbf{u}_l\right)\right]|^2}, \; l = 1, 2, \ldots, L.
\end{aligned}
$$

(10.124)

Hence, $h_{\mathrm{T},l}$ is a transform-domain Wiener filter with adjustable input noise level $\mu_l \phi_{vv}(\mathbf{u}_l)$. Since $0 \leq h_{\mathrm{T},l} \leq 1$, $\forall \mu_l \geq 0$, then $\mathrm{oSNR}\left(\mathbf{h}_{\mathrm{T}}, \mathbf{U}\right) \geq \mathrm{iSNR}(\mathbf{U})$. Therefore, this method improves the (fullband) output SNR.

The Lagrange multiplier must satisfy

$$J_v\left(h_{\mathrm{T},l}, \mathbf{u}_l\right) = \beta_l \phi_v(\mathbf{u}_l) = h_{\mathrm{T},l}^2 \phi_v(\mathbf{u}_l), \; l = 1, 2, \ldots, L. \quad (10.125)$$

Substituting (10.124) into (10.125), we can find

$$\mu_l = \mathrm{iSNR}(\mathbf{u}_l)\left(\frac{1}{\sqrt{\beta_l}} - 1\right), \; l = 1, 2, \ldots, L \quad (10.126)$$

and from (10.125), we also have

$$h_{\mathrm{T},l} = \sqrt{\beta_l}, \; l = 1, 2, \ldots, L. \quad (10.127)$$

The Lagrange multiplier μ_l can always be chosen in an ad-hoc way if we prefer. Then, we can see from (10.124) that there are four cases:

- $\mu_l = 1$; in this case, the tradeoff and Wiener filters are the same, i.e., $h_{\mathrm{T},l} = h_{\mathrm{W},l}$;
- $\mu_l = 0$; in this circumstance, we have $h_{\mathrm{T},l} = 1$ and there is no noise reduction and no speech distortion;
- $\mu_l > 1$; this situation corresponds to a more aggressive noise reduction at the expense of higher speech distortion as compared to the Wiener filter;
- $\mu_l < 1$; this case corresponds to a more conservative noise reduction with less speech distortion as compared to the Wiener filter.

With (10.23) and (10.127) we can rewrite, equivalently, the transform-domain tradeoff filter into the time domain:

$$\mathbf{H}_{\mathrm{T}}(\mathbf{U}) = \mathbf{R}_y^{1/2} \mathbf{U} \, \mathrm{diag}^{1/2}\left[\beta_1 \; \beta_2 \; \cdots \; \beta_L\right] \mathbf{U}^H \mathbf{R}_y^{-1/2}. \quad (10.128)$$

10.5.4 Examples of Unitary Matrices

There are perhaps a very large number of unitary (or orthogonal) matrices that can be used in tandem with the different noise reduction filters presented in this section. But does a transformation exist in such a way that an optimal filter maximizes noise reduction while it minimizes speech distortion at the same time? The answer to this question is not straightforward. However, intuitively we believe that some unitary matrices will be more effective than others for a given noise reduction filter.

The first obvious choice is the KLT developed in Chapter 2. In this case, $\mathbf{U} = \mathbf{Q}_y$ where $\mathbf{Q}_y$ contains the eigenvectors of the correlation matrix, $\mathbf{R}_y$, of the noisy signal, $y(k)$, for which the spectral representation are the eigenvalues of $\mathbf{R}_y$. This choice seems to be the most natural one since the Parseval's theorem is verified.

Another choice for $\mathbf{U}$ is the Fourier matrix:

$$\mathbf{F} = \begin{bmatrix} \mathbf{f}_1 \ \mathbf{f}_2 \ \cdots \ \mathbf{f}_L \end{bmatrix}, \tag{10.129}$$

where

$$\mathbf{f}_l = \frac{1}{\sqrt{L}} \begin{bmatrix} 1 \ \exp(j\omega_l) \ \cdots \ \exp[j\omega_l(L-1)] \end{bmatrix}^T \tag{10.130}$$

and $\omega_l = 2\pi(l-1)/L$, $l = 1, \ldots, L$. Even though $\mathbf{F}$ is unitary, the matrix $\mathbf{G}(\mathbf{F})$ constructed from $\mathbf{F}$ is not; as a result, the Parseval's theorem does not hold but the transform signals at the different frequencies are uncorrelated. Filters in this new Fourier domain will probably perform differently as compared to the classical frequency-domain filters.

In our application, all signals are real and it may be more convenient to select an orthogonal matrix instead of a unitary one. So another choice close to the previous one is the discrete cosine transform:

$$\mathbf{C} = \begin{bmatrix} \mathbf{c}_1 \ \mathbf{c}_2 \ \cdots \ \mathbf{c}_L \end{bmatrix}, \tag{10.131}$$

where

$$\mathbf{c}_l = \begin{bmatrix} c(1) \ c(2) \cos \frac{\pi(2l-1)}{2L} \ \cdots \ c(L) \cos \frac{\pi(2l-1)(L-1)}{2L} \end{bmatrix}^T \tag{10.132}$$

with $c(1) = \sqrt{1/L}$ and $c(l) = \sqrt{2/L}$ for $l \neq 1$. We can verify that $\mathbf{C}^T\mathbf{C} = \mathbf{C}\mathbf{C}^T = \mathbf{I}$.

Another possible choice for $\mathbf{U}$ is the Hadamard transform. We are now going to give an example of a Hadamard matrix. Let us define first the two matrices:

$$\mathbf{H}_1 = 1, \tag{10.133}$$

$$\mathbf{H}_2 = \frac{1}{\sqrt{2}} \begin{bmatrix} 1 & 1 \\ 1 & -1 \end{bmatrix} = \frac{1}{\sqrt{2}} \mathbf{H}_2'. \tag{10.134}$$

146   10 Optimal Filters in the Transform Domain

A Hadamard matrix of order 2^p can be constructed from the two previous matrices as follows:

$$\mathbf{H}_{2^p} = \frac{1}{\sqrt{2^p}} \begin{bmatrix} \mathbf{H}'_{2^{p-1}} & \mathbf{H}'_{2^{p-1}} \\ \mathbf{H}'_{2^{p-1}} & -\mathbf{H}'_{2^{p-1}} \end{bmatrix}$$
$$= \frac{1}{\sqrt{2^p}} \mathbf{H}'_2 \otimes \mathbf{H}'_{2^{p-1}} \qquad (10.135)$$

for $p \geq 2$, where $\otimes$ is the Kronecker product. It can be verified that $\mathbf{H}_{2^p} \mathbf{H}_{2^p}^T = \mathbf{H}_{2^p}^T \mathbf{H}_{2^p} = \mathbf{I}$.

10.6 Experiments

We have formulated the noise reduction problem in a generalized transform domain and discussed the design of different noise reduction filters in that domain. In this section, we study several filters through experiments and compare different transforms for their impact on noise reduction performance. The experimental setup is the same as used in Chapter 7.

10.6.1 Performance of Wiener Filter in White Gaussian Noise

In this experiment, we examine the performance of the Wiener filter in a white Gaussian noise with an iSNR = 10 dB.

Similar to the time- and KLE-domain filters, the implementation of the generalized transform-domain Wiener filter requires to know the correlation matrices $\mathbf{R}_y$ and $\mathbf{R}_v$. Although different approaches can be used to compute $\mathbf{R}_y$ from the noisy signal $y(k)$, we will adopt the recursive method given in (7.44). Same as in Section 7.4, we do not to use any noise estimator, but compute the noise correlation matrix $\mathbf{R}_v$ directly from the noise signal. Since the background noise is stationary, we use a long-term sample average to compute $\mathbf{R}_v$.

With the above ways of computing $\mathbf{R}_y$ and $\mathbf{R}_v$, the performance of the Wiener filter given in (10.92) is mainly affected by three major elements: the forgetting factor α_y, the frame length L, and the transformation matrix $\mathbf{U}$. In the first experiment, we fix the frame length to $L = 32$ and study the effect of the forgetting factor α_y with different transforms on the performance. For the matrix $\mathbf{U}$, we choose to compare four well-known transforms: KL, Fourier, cosine, and Hadamard.

Figure 10.1 plots the output SNR and the speech-distortion index for different transforms as a function of the forgetting factor α_y. It is seen that the output SNR for all the studied transforms first increases as α_y increases and then decreases. The highest output SNR is obtained when α_y is between 0.985 and 0.995. This agrees with what has been observed with the time- and KLE-domain filters.

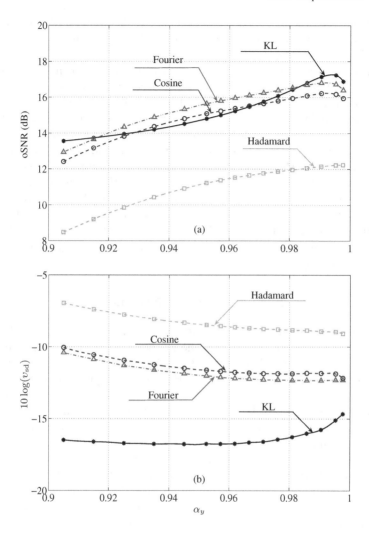

Fig. 10.1. Noise reduction performance of the Wiener filter versus α_y in white Gaussian noise: iSNR = 10 dB and $L = 32$.

Generally, the speech-distortion index v_{sd} decreases as the forgetting factor α_y increases. So, the larger the value of α_y, the smaller the speech distortion index. The only exception is with the KL transform, where the speech-distortion index first decreases slightly and then increases as α_y increases from 0.9 to its upper bound 1.

We also see from Fig. 10.1 that the Fourier and cosine transforms yielded similar performance. When α_y is reasonably large (e.g., ≥ 0.98), the KL, Fourier, and cosine transforms produced similar output SNRs. Comparatively, however, the KL transform has much lower speech distortion. When the value

148   10 Optimal Filters in the Transform Domain

of α_y is in a reasonable range (e.g., $\in [0.96, 1]$), the Hadamard transform can also improve the SNR. But its performance is much inferior to that of the other three transforms.

The transform-domain Wiener filter with the KL transform is not identical to the KLE-domain Wiener filter (Class I) derived in Chapter 9. Indeed, it is noticed from Fig. 10.1 that the speech-distortion index with the KL transform is somehow slightly different from what was observed with the KLE-domain Wiener filter in Fig. 9.1. There are three major reasons for this difference: 1) in Chapter 9, the matrix $\mathbf{U}$ is obtained by eigenvalue decomposition of the correlation matrix $\mathbf{R}_x$, while the matrix $\mathbf{U}$ in this chapter is derived from the correlation matrix $\mathbf{R}_y$; 2) the frame length L in Fig. 9.1 is 20 while it is 32 in this experiment (because we also used the Hadamard transform for which the value of L has to be a power of 2); 3) in this section, a small positive constant has been applied to regularize the correlation matrix $\mathbf{R}_y$ so that the inverse of its square root, i.e., $\mathbf{R}_y^{-1/2}$, can be computed reliably, while we do not regularize any matrix in Chapter 9.

10.6.2 Effect of Filter Length on Performance

In this experiment, we study the effect of the frame length L on the noise reduction performance. Same as in the previous experiment, white Gaussian noise is used with an iSNR = 10 dB. Again, the noise correlation matrix is computed using a long-term average. Based on the previous results, we set $\alpha_y = 0.99$. Figure 10.2 depicts the output SNR and speech-distortion index, both as a function of L. It is seen that, as L increases, the output SNR of the Wiener filter using the KL transform first increases and then decreases. Good performance with this transform is obtained when L is between 20 and 80. The Fourier and cosine transforms yielded similar performance. For the Hadamard transform, a larger L corresponds to a less SNR gain and a larger speech-distortion index, which indicates that a small frame length L should be preferred if the Hadamard transform is used. Generally, however, the Hadamard transform is much inferior to the KL, Fourier, and cosine transforms in performance.

10.6.3 Performance of Tradeoff Filter in White Gaussian Noise

Now we evaluate the performance of the transform-domain tradeoff filter given in (10.124) in different conditions. From the theoretical analysis, we already know that if $\mu = 1$, the tradeoff filter is the Wiener filter. Increasing the value of μ will give more noise reduction, but will also lead to more speech distortion. In this experiment, we set $\mu = 4$. Again, the noise used is a white Gaussian random process with an iSNR = 10 dB. The noise correlation matrix is computed using a long-term average. We first fix the frame length L to 32 and investigate the effect of α_y and different transforms on the performance.

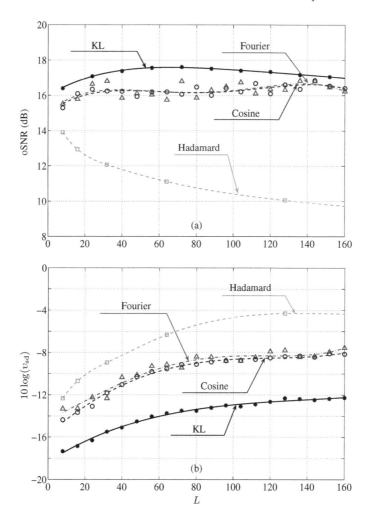

Fig. 10.2. Noise reduction performance of the Wiener filter versus L in white Gaussian noise: iSNR = 10 dB and $\alpha_y = 0.99$.

Figure 10.3 portrays the output SNR and speech-distortion index as a function of α_y.

Similar to the Wiener filter case, the output SNR (for all the studied transforms) first increases and then drops as α_y increases. The largest SNR gain for each transform is obtained when α_y is between 0.985 and 0.995. The KL transform yielded the best performance (with the highest output SNR and lowest speech-distortion index). The Fourier and cosine transforms behave similarly. In general, the performance of the Hadamard transform is

150   10 Optimal Filters in the Transform Domain

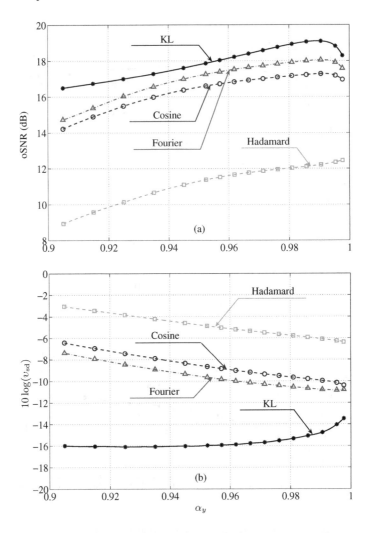

Fig. 10.3. Noise reduction performance of the tradeoff filter versus α_y in white Gaussian noise: iSNR = 10 dB, $L = 32$, and $\mu = 4$.

relatively poor as compared to the other three transforms, again, indicating that this transform is less effective for the purpose of noise reduction.

Comparing Figs. 10.3 and 10.1, one can see that the output SNR of the tradeoff filter is boosted with the use of a larger value of μ, but this is achieved at the price of adding more speech distortion, which corroborates the analysis presented in Section 10.5.3.

To investigate the effect of the frame length L on the performance, we set $\alpha_y = 0.99$ and change L from 4 to 160 (note that, for the Hadamard transform, the length L can only take values that are powers of 2). All other

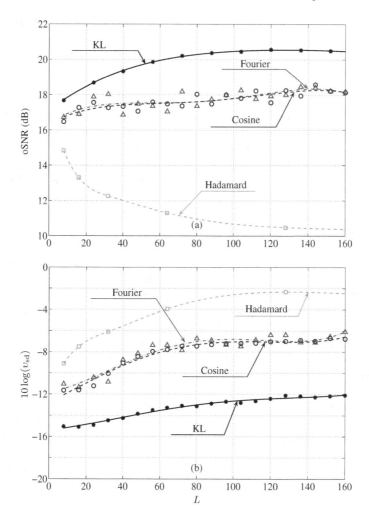

Fig. 10.4. Noise reduction performance of the tradeoff filter versus L in white Gaussian noise: iSNR = 10 dB, $\alpha_y = 0.99$, and $\mu = 4$.

conditions are the same as used in the previous experiment. The results are shown in Fig. 10.4. Similar to the Wiener filter case, we observe that the output SNR for the KL transform first increases to its maximum and then drops as L increases. However, there are two major differences as compared to the Wiener filter case: 1) the largest SNR gain with the tradeoff filter appears when L is in the range of 40−120, while such performance occurs when L in the range of 40−80 for the Wiener filter; 2) although the performance with the KL transform decreases if we keep increasing L after the optimal performance is achieved, the performance degradation with L is almost negligible. The reason

152   10 Optimal Filters in the Transform Domain

for these two differences can be explained as follows. In this experiment, we set $\mu = 4$ and all the β_l in the diagonal matrix $\text{diag}^{1/2}[\beta_1 \quad \beta_2 \quad \cdots \quad \beta_L]$ that are less than 0 are forced to 0. After a certain threshold, if we further increase L, the signal subspace that consists of all the positive β_l value do not increase much. In other words, even though we increases L, which results in a larger size for $\mathbf{R}_y$, we are still dealing with a signal subspace of similar order. As a result, the performance does not change much. Again, the Fourier and cosine transforms have similar performance. Comparatively, the effect of L on the Fourier, cosine, and Hadamard transforms in the subspace case is almost the same as that in the Wiener filter situation. The only difference is that now we have achieved a higher SNR gain. But the speech distortion is also higher.

10.7 Summary

In this chapter, we formulated the noise reduction problem in a generalized transform domain, where any unitary (or orthogonal) matrix can be used to construct the forward (for analysis) and inverse (for synthesis) transforms. The advantages of working in this generalized domain are multiple, such as different transforms can be used to replace each other without any requirement to change the algorithm formulation (optimal filter) and it is easier to fairly compare different transforms for their noise reduction performance. We have addressed the design of different optimal and suboptimal filters in such a generalized transform domain, including the Wiener filter, the parametric Wiener filter, the tradeoff filter, etc. We have also compared, through experiments, four different transforms (KL, Fourier, cosine, and Hadamard) for their noise reduction performance. In general, the KL transform yielded the best performance. The Fourier and cosine transforms have quite similar performance, which is slightly poorer than that of the KL transform. While the Hadamard transform can improve the SNR, it is much less effective than the other three transforms for noise reduction.

11

Spectral Enhancement Methods

In this chapter, we focus on spectral enhancement methods in the short-time Fourier transform (STFT) domain. We present statistical models for the speech and noise signals, and derive estimators for the speech signal using various distortion measures. We also describe an example of a speech enhancement algorithm and demonstrate its performance on speech signals degraded by various additive noise types.

11.1 Problem Formulation

In the STFT domain, (2.1) can be rewritten as

$$Y(n, m) = X(n, m) + V(n, m), \tag{11.1}$$

where $Y(n, m)$, $X(n, m)$, and $V(n, m)$ are respectively the STFTs of $y(k)$, $x(k)$, and $v(k)$, at time-frame n and frequency-bin m. We recall that the STFT and the inverse transform are

$$A(n, m) = \sum_{k=0}^{M-1} a(k + nL)\, \psi(k)\, e^{-j\frac{2\pi}{M} km}, \tag{11.2}$$

$$a(k) = \sum_{n} \sum_{m=0}^{M-1} A(n, m)\, \tilde{\psi}(k - nL)\, e^{j\frac{2\pi}{N} m(k - nL)}, \tag{11.3}$$

where $A \in \{Y, X, V\}$, $a \in \{y, x, v\}$, $\psi(k)$ is an analysis window of size M (e.g., Hamming window), L is the framing step (number of samples separating two successive time frames), and $\tilde{\psi}(k)$ is a synthesis window that is biorthogonal to the analysis window $\psi(k)$ [124]. Another possible form for (11.1) is

$$A_y(n, m)e^{j\varphi_y(n,m)} = A_x(n, m)e^{j\varphi_x(n,m)} + A_v(n, m)e^{j\varphi_v(n,m)}, \tag{11.4}$$

J. Benesty et al., *Noise Reduction in Speech Processing*, Springer Topics in Signal Processing 2,
DOI 10.1007/978-3-642-00296-0_11, © Springer-Verlag Berlin Heidelberg 2009

154   11 Spectral Enhancement Methods

where for any random signal $A(n,m) = A_a(n,m)e^{j\varphi_a(n,m)}$, $A_a(n,m)$ and $\varphi_a(n,m)$ are its amplitude and phase at time-frame n and frequency-bin $m = 0, \ldots, M-1$, $a \in \{y, x, v\}$.

Using the variances of the noisy spectral coefficients and the fact that $x(k)$ and $v(k)$ are uncorrelated, we get

$$\phi_y(n,m) = \phi_x(n,m) + \phi_v(n,m), \tag{11.5}$$

where

$$\phi_a(n,m) = E\left[|A(n,m)|^2\right]$$
$$= E\left[A_a^2(n,m)\right]$$

is the variance of a spectral coefficient of the signal $a(k)$ at time-frame n and frequency-bin m.

An estimate of $X(n,m)$ can be obtained by multiplying $Y(n,m)$ with a complex gain, i.e.,

$$Z(n,m) = H(n,m)Y(n,m) \tag{11.6}$$
$$= H(n,m)\left[X(n,m) + V(n,m)\right]$$
$$= X_{\mathrm{F}}(n,m) + V_{\mathrm{F}}(n,m),$$

where $Z(n,m)$ is the STFT of the signal $z(k)$. The variance of a spectral coefficient of $z(k)$ is then

$$\phi_z(n,m) = |H(n,m)|^2\, \phi_y(n,m)$$
$$= |H(n,m)|^2\, [\phi_x(n,m) + \phi_v(n,m)]. \tag{11.7}$$

We can go back to the time domain to obtain the estimate

$$z(k) = \sum_n \sum_{m=0}^{M-1} Z(n,m)\, \tilde{\psi}(k-nL)\, e^{j\frac{2\pi}{N}m(k-nL)}. \tag{11.8}$$

The spectral enhancement problem is generally formulated as deriving an estimator $Z(n,m)$ for the speech spectral coefficients, such that the expected value of a certain distortion measure is minimized. Let $d[X(n,m), Z(n,m)]$ denote a distortion measure between $X(n,m)$ and its estimate $Z(n,m)$, let $\hat{\phi}_x(n,m)$ denote an estimate for the variance of a speech spectral coefficient $X(n,m)$, and let $\hat{\phi}_v(n,m)$ denote an estimate for the variance of a noise spectral coefficient $V(n,m)$. Then an estimator for $X(n,m)$ which minimizes the expected distortion given $\hat{\phi}_x(n,m)$, $\hat{\phi}_v(n,m)$, and the noisy spectral coefficient $Y(n,m)$ is obtained by solving the minimization problem

$$\min_{Z(n,m)} E\left\{d[X(n,m), Z(n,m)] \,\Big|\, \hat{\phi}_x(n,m),\, \hat{\phi}_v(n,m),\, Y(n,m)\right\}. \tag{11.9}$$

In particular, restricting ourselves to a squared error distortion measure of the form

$$d\left[X(n,m), Z(n,m)\right] = |g[Z(n,m)] - g[X(n,m)]|^2, \qquad (11.10)$$

where $g(X)$ is a specific function of X (e.g., X, $|X|$, $\log|X|$, $e^{j\angle X}$), the estimator $Z(n,m)$ can be calculated from

$$g[Z(n,m)] = E\left\{ g[X(n,m)] \mid \hat{\phi}_x(n,m), \hat{\phi}_v(n,m), Y(n,m) \right\}. \qquad (11.11)$$

The design of a particular spectral enhancement algorithm requires to specify the function $g(X)$ that determines the fidelity criterion of the estimator, and estimators $\hat{\phi}_x(n,m)$ and $\hat{\phi}_v(n,m)$ for the speech and noise spectral variances.

11.2 Performance Measures

11.2.1 SNR

The input SNR in the STFT domain is defined as the ratio of the intensity of the signal of interest (speech) over the intensity of the background noise, i.e.,

$$\text{iSNR} = \frac{\sum_n \sum_{m=0}^{M-1} \phi_x(n,m)}{\sum_n \sum_{m=0}^{M-1} \phi_v(n,m)}. \qquad (11.12)$$

We also define the subband input SNR and segmental input SNR in the STFT domain as

$$\text{iSNR}(n,m) = \frac{\phi_x(n,m)}{\phi_v(n,m)}, \ m = 0, 1, \ldots, M-1, \qquad (11.13)$$

$$\text{iSNR}(n) = \frac{\sum_{m=0}^{M-1} \phi_x(n,m)}{\sum_{m=0}^{M-1} \phi_v(n,m)}. \qquad (11.14)$$

After noise reduction with the STFT domain model given in (11.6), the subband, segmental, and fullband output SNRs are

$$\text{oSNR}\left[H(n,m)\right] = \frac{|H(n,m)|^2 \phi_x(n,m)}{|H(n,m)|^2 \phi_v(n,m)} \qquad (11.15)$$
$$= \text{iSNR}(n,m), \ m = 0, 1, \ldots, M-1,$$

$$\text{oSNR}\left[H(n)\right] = \frac{\sum_{m=0}^{M-1} |H(n,m)|^2 \phi_x(n,m)}{\sum_{m=0}^{M-1} |H(n,m)|^2 \phi_v(n,m)}, \qquad (11.16)$$

156   11 Spectral Enhancement Methods

$$\text{oSNR}(H) = \frac{\sum_n \sum_{m=0}^{M-1} |H(n,m)|^2 \, \phi_x(n,m)}{\sum_n \sum_{m=0}^{M-1} |H(n,m)|^2 \, \phi_v(n,m)}. \tag{11.17}$$

It is important to find a complex gain $H(n,m)$, $n \in \mathbb{Z}$, $m = 0,1,\ldots,M-1$, in such a way that $\text{oSNR}(H) \geq \text{iSNR}$.

Property 11.1. We always have

$$\sum_{m=0}^{M-1} \text{iSNR}(n,m) \geq \text{iSNR}(n), \tag{11.18}$$

$$\sum_n \sum_{m=0}^{M-1} \text{iSNR}(n,m) \geq \sum_n \text{iSNR}(n) \geq \text{iSNR}, \tag{11.19}$$

$$\sum_{m=0}^{M-1} \text{oSNR}\left[H(n,m)\right] \geq \text{oSNR}\left[H(n)\right], \tag{11.20}$$

$$\sum_n \sum_{m=0}^{M-1} \text{oSNR}\left[H(n,m)\right] \geq \sum_n \text{oSNR}\left[H(n)\right] \geq \text{oSNR}(H). \tag{11.21}$$

This means that the aggregation of the subband SNRs is greater than or equal to the segmental SNR and the aggregation of the segmental SNRs is greater than or equal to the fullband SNR.

Proof. Easy to show by using inequality (3.16).

11.2.2 Noise-Reduction Factor

By analogy to the time-domain and frequency-domain definitions, we define the subband, segmental, and fullband noise-reduction factors in the STFT domain as

$$\xi_{\text{nr}}\left[H(n,m)\right] = \frac{\phi_v(n,m)}{|H(n,m)|^2 \, \phi_v(n,m)}$$

$$= \frac{1}{|H(n,m)|^2}, \quad m = 0,1,\ldots,M-1, \tag{11.22}$$

$$\xi_{\text{nr}}\left[H(n)\right] = \frac{\sum_{m=0}^{M-1} \phi_v(n,m)}{\sum_{m=0}^{M-1} |H(n,m)|^2 \, \phi_v(n,m)}, \tag{11.23}$$

$$\xi_{\text{nr}}(H) = \frac{\sum_n \sum_{m=0}^{M-1} \phi_v(n,m)}{\sum_n \sum_{m=0}^{M-1} |H(n,m)|^2 \, \phi_v(n,m)}. \tag{11.24}$$

The segmental and fullband noise-reduction factors are ratios of the noise energy over a weighted energy of the noise with the weighting $\xi_{\text{nr}}^{-1}\left[H(n,m)\right]$. After the filtering operation, the residual noise level at time-frame n and frequency-bin m is expected to be lower than that of the original noise level, therefore $\xi_{\text{nr}}\left[H(n,m)\right]$, $\xi_{\text{nr}}\left[H(n)\right]$, and $\xi_{\text{nr}}(H)$ are expected to larger than 1.

Property 11.2. We always have

$$\sum_{m=0}^{M-1} \xi_{\mathrm{nr}}\left[H(n,m)\right] \geq \xi_{\mathrm{nr}}\left[H(n)\right], \tag{11.25}$$

$$\sum_{n}\sum_{m=0}^{M-1} \xi_{\mathrm{nr}}\left[H(n,m)\right] \geq \sum_{n}\xi_{\mathrm{nr}}\left[H(n)\right] \geq \xi_{\mathrm{nr}}(H). \tag{11.26}$$

This means that the aggregation of the subband noise-reduction factors is greater than or equal to the segmental noise-reduction factor and the aggregation of the segmental noise-reduction factors is greater than or equal to the fullband noise-reduction factor.

Proof. Easy to show by using inequality (3.16).

11.2.3 Speech-Distortion Index

The definition of the speech-distortion index in the frequency domain is extended to the STFT domain. We define the subband speech-distortion index in the STFT domain as

$$\upsilon_{\mathrm{sd}}\left[H(n,m)\right] = \frac{E\left[|H(n,m)X(n,m) - X(n,m)|^2\right]}{\phi_x(n,m)}$$
$$= |1 - H(n,m)|^2, \; m = 0,1,\ldots,M-1. \tag{11.27}$$

The subband speech-distortion index is lower bounded by 0 and expected to be lower than 1 for optimal filters. The higher the value of $\upsilon_{\mathrm{sd}}\left[H(n,m)\right]$, the more the speech is distorted at time-frame n and frequency-bin m.

The segmental and fullband speech-distortion indices in the STFT domain are

$$\upsilon_{\mathrm{sd}}\left[H(n)\right] = \frac{\sum_{m=0}^{M-1} E\left[|H(n,m)X(n,m) - X(n,m)|^2\right]}{\sum_{m=0}^{M-1} \phi_x(n,m)}$$
$$= \frac{\sum_{m=0}^{M-1} |1 - H(n,m)|^2 \phi_x(n,m)}{\sum_{m=0}^{M-1} \phi_x(n,m)}$$
$$= \frac{\sum_{m=0}^{M-1} \upsilon_{\mathrm{sd}}\left[H(n,m)\right] \phi_x(n,m)}{\sum_{m=0}^{M-1} \phi_x(n,m)} \tag{11.28}$$

and

$$\upsilon_{\mathrm{sd}}(H) = \frac{\sum_{n}\sum_{m=0}^{M-1} E\left[|H(n,m)X(n,m) - X(n,m)|^2\right]}{\sum_{n}\sum_{m=0}^{M-1} \phi_x(n,m)}$$
$$= \frac{\sum_{n}\sum_{m=0}^{M-1} \upsilon_{\mathrm{sd}}\left[H(n,m)\right] \phi_x(n,m)}{\sum_{n}\sum_{m=0}^{M-1} \phi_x(n,m)}. \tag{11.29}$$

158   11 Spectral Enhancement Methods

Equations (11.28) and (11.29) are ratios of the weighted energy of the speech with the weighting $v_{\mathrm{sd}}[H(n,m)]$ over the energy of the speech. If $v_{\mathrm{sd}}[H(n,m)] \leq 1$, $\forall(n,m)$, we see from (11.28) and (11.29) that $v_{\mathrm{sd}}[H(n)] \leq 1$ and $v_{\mathrm{sd}}(H) \leq 1$.

Property 11.3. We always have

$$\sum_{m=0}^{M-1} v_{\mathrm{sd}}[H(n,m)] \geq v_{\mathrm{sd}}[H(n)], \tag{11.30}$$

$$\sum_{n} \sum_{m=0}^{M-1} v_{\mathrm{sd}}[H(n,m)] \geq \sum_{n} v_{\mathrm{sd}}[H(n)] \geq v_{\mathrm{sd}}(H). \tag{11.31}$$

This means that the aggregation of the subband speech-distortion indices is greater than or equal to the segmental speech-distortion index and the aggregation of the segmental speech-distortion indices is greater than or equal to the fullband speech-distortion index.

Proof. Easy to show by using inequality (3.16).

11.2.4 Speech-Reduction Factor

We define the subband speech-reduction factor in the STFT domain as

$$\xi_{\mathrm{sr}}[H(n,m)] = \frac{\phi_x(n,m)}{|H(n,m)|^2 \phi_x(n,m)}$$

$$= \frac{1}{|H(n,m)|^2}, \quad m = 0, 1, \ldots, M - 1. \tag{11.32}$$

The larger the value of $\xi_{\mathrm{sr}}[H(n,m)]$, the more the speech is reduced at time-frame n and frequency-bin m. The segmental and fullband speech-reduction factors in the STFT domain are

$$\xi_{\mathrm{sr}}[H(n)] = \frac{\sum_{m=0}^{M-1} \phi_x(n,m)}{\sum_{m=0}^{M-1} |H(n,m)|^2 \phi_x(n,m)} \tag{11.33}$$

$$= \frac{\sum_{m=0}^{M-1} \phi_x(n,m)}{\sum_{m=0}^{M-1} \xi_{\mathrm{sr}}^{-1}[H(n,m)] \phi_x(n,m)},$$

$$\xi_{\mathrm{sr}}(H) = \frac{\sum_{n} \sum_{m=0}^{M-1} \phi_x(n,m)}{\sum_{n} \sum_{m=0}^{M-1} |H(n,m)|^2 \phi_x(n,m)} \tag{11.34}$$

$$= \frac{\sum_{n} \sum_{m=0}^{M-1} \phi_x(n,m)}{\sum_{n} \sum_{m=0}^{M-1} \xi_{\mathrm{sr}}^{-1}[H(n,m)] \phi_x(n,m)}.$$

The segmental and fullband speech-reduction factors in the STFT domain are ratios of the energy of the speech over the weighted energy of the speech with the weighting $\xi_{\mathrm{sr}}^{-1}[H(n,m)]$.

Property 11.4. We always have

$$\sum_{m=0}^{M-1} \xi_{sr}\left[H(n,m)\right] \geq \xi_{sr}\left[H(n)\right], \tag{11.35}$$

$$\sum_{n}\sum_{m=0}^{M-1} \xi_{sr}\left[H(n,m)\right] \geq \sum_{n} \xi_{sr}\left[H(n)\right] \geq \xi_{sr}(H). \tag{11.36}$$

This means that the aggregation of the subband speech-reduction factors is greater than or equal to the segmental speech-reduction factor and the aggregation of the segmental speech-reduction factors is greater than or equal to the fullband speech-reduction factor.

Proof. Easy to show by using inequality (3.16).

Property 11.5. The fullband and segmental measures mentioned above are related in the STFT domain by

$$\frac{\text{oSNR}(H)}{\text{iSNR}} = \frac{\xi_{nr}(H)}{\xi_{sr}(H)}, \tag{11.37}$$

$$\frac{\text{oSNR}\left[H(n)\right]}{\text{iSNR}(n)} = \frac{\xi_{nr}\left[H(n)\right]}{\xi_{sr}\left[H(n)\right]}. \tag{11.38}$$

Proof. This is easy to see by combining expressions (11.12), (11.17), (11.24), and (11.34), and combining expressions (11.14), (11.16), (11.23), and (11.33).

Hence, a segmental or fullband increase in the SNR, i.e., $\text{oSNR}\left[H(n)\right] > \text{iSNR}(n)$ or $\text{oSNR}(H) > \text{iSNR}$, can be obtained only when the segmental, or respectively the fullband, noise-reduction factor is larger than the corresponding speech-reduction factor, i.e., $\xi_{nr}\left[H(n)\right] > \xi_{sr}\left[H(n)\right]$ or $\xi_{nr}(H) > \xi_{sr}(H)$.

11.3 MSE Criterion

We define the error signal between the estimated and desired signals at time-frame n and frequency-bin m as

$$\begin{aligned}\mathcal{E}(n,m) &= Z(n,m) - X(n,m) \\ &= H(n,m)Y(n,m) - X(n,m).\end{aligned} \tag{11.39}$$

This error can also be put into the form:

$$\mathcal{E}(n,m) = \mathcal{E}_x(n,m) + \mathcal{E}_v(n,m), \tag{11.40}$$

where

$$\mathcal{E}_x(n,m) = \left[H(n,m) - 1\right]X(n,m) \tag{11.41}$$

160   11 Spectral Enhancement Methods

is the speech distortion due to the complex gain, and

$$\mathcal{E}_v(n,m) = H(n,m)V(n,m) \qquad (11.42)$$

represents the residual noise.

The subband MSE in the STFT domain is then

$$J[H(n,m)] = E\left[|\mathcal{E}(n,m)|^2\right] \qquad (11.43)$$
$$= E\left[|\mathcal{E}_x(n,m)|^2\right] + E\left[|\mathcal{E}_v(n,m)|^2\right]$$
$$= |1 - H(n,m)|^2 \phi_x(n,m) + |H(n,m)|^2 \phi_v(n,m)$$
$$= J_x[H(n,m)] + J_v[H(n,m)].$$

For the particular gain $H(n,m) = 1$, $\forall(n,m)$, we get

$$J(1) = \phi_v(n,m), \qquad (11.44)$$

so there will be neither noise reduction nor speech distortion. Therefore, we define the subband NMSE in the STFT domain as

$$\tilde{J}[H(n,m)] = \frac{J[H(n,m)]}{J(1)}$$
$$= \text{iSNR}(n,m) \cdot \upsilon_{\text{sd}}[H(n,m)] + \frac{1}{\xi_{\text{nr}}[H(n,m)]}, \qquad (11.45)$$

where

$$\upsilon_{\text{sd}}[H(n,m)] = \frac{J_x[H(n,m)]}{\phi_x(n,m)}, \qquad (11.46)$$
$$\xi_{\text{nr}}[H(n,m)] = \frac{\phi_v(n,m)}{J_v[H(n,m)]}. \qquad (11.47)$$

The STFT-domain NMSE depends explicitly on the subband speech-distortion index and the subband noise-reduction factor.

We define the segmental and fullband MSEs as

$$J[H(n)] = \frac{1}{M} \sum_{m=0}^{M-1} J[H(n,m)] \qquad (11.48)$$
$$= \frac{1}{M} \sum_{m=0}^{M-1} \phi_x(n,m) |1 - H(n,m)|^2 + \frac{1}{M} \sum_{m=0}^{M-1} \phi_v(n,m) |H(n,m)|^2$$
$$= J_x[H(n)] + J_v[H(n)],$$

$$J(H) = \frac{1}{NM} \sum_{n=0}^{N-1} \sum_{m=0}^{M-1} J[H(n,m)] \tag{11.49}$$

$$= \frac{1}{NM} \sum_{n=0}^{N-1} \sum_{m=0}^{M-1} \phi_x(n,m) |1 - H(n,m)|^2 +$$

$$\frac{1}{NM} \sum_{n=0}^{N-1} \sum_{m=0}^{M-1} \phi_v(n,m) |H(n,m)|^2$$

$$= J_x(H) + J_v(H).$$

The segmental and fullband normalized MSEs are

$$\tilde{J}[H(n)] = M \frac{J[H(n)]}{\sum_{m=0}^{M-1} \phi_v(n,m)} \tag{11.50}$$

$$= \frac{\sum_{m=0}^{M-1} \phi_x(n,m) |1 - H(n,m)|^2}{\sum_{m=0}^{M-1} \phi_v(n,m)} + \frac{\sum_{m=0}^{M-1} \phi_v(n,m) |H(n,m)|^2}{\sum_{m=0}^{M-1} \phi_v(n,m)}$$

$$= \text{iSNR}(n) \cdot \upsilon_{\text{sd}}[H(n)] + \frac{1}{\xi_{\text{nr}}[H(n)]},$$

$$\tilde{J}(H) = NM \frac{J(H)}{\sum_{n=0}^{N-1} \sum_{n=0}^{N-1} \sum_{m=0}^{M-1} \phi_v(n,m)} \tag{11.51}$$

$$= \frac{\sum_{n=0}^{N-1} \sum_{m=0}^{M-1} \phi_x(n,m) |1 - H(n,m)|^2}{\sum_{m=0}^{M-1} \phi_v(n,m)} +$$

$$\frac{\sum_{n=0}^{N-1} \sum_{m=0}^{M-1} \phi_v(n,m) |H(n,m)|^2}{\sum_{m=0}^{M-1} \phi_v(n,m)}$$

$$= \text{iSNR} \cdot \upsilon_{\text{sd}}(H) + \frac{1}{\xi_{\text{nr}}(H)}.$$

Same as in the time, frequency, and KLE domains, in the STFT domain the NMSE (subband, segmental, or fullband) depends explicitly on the input SNR, speech-distortion index, and noise-reduction factor.

11.4 Signal Model

In this section, we present a Gaussian statistical model that takes into account the time-correlation between successive spectral components of the speech signal. It is commonly assumed that expansion coefficients in different frequency-bins are statistically independent [31], [32], [44], [90]. This allows to formulate independent estimation problems for each frequency-bin m, which greatly simplifies the resulting algorithms. A Gaussian statistical model in the STFT domain relies on the following assumptions [30]:

162   11 Spectral Enhancement Methods

1. The noise spectral coefficients $\{V(n,m)\}$ are zero-mean statistically independent Gaussian random variables. The real and imaginary parts of $V(n,m)$ are independent and identically distributed (iid) random variables $\sim \mathcal{N}\left[0, \frac{1}{2}\phi_v(n,m)\right]$.
2. Given $\{\phi_x(n,m)\}$, the speech spectral coefficients $\{X(n,m)\}$ are generated by

$$X(n,m) = \sqrt{\phi_x(n,m)}\, W(n,m), \qquad (11.52)$$

where $\{W(n,m)\}$ is a white Gaussian process with zero mean $(E\left[W(n,m)\right] = 0)$ and unit variance $(E\left[|W(n,m)|^2\right] = 1)$. The probability density function (pdf) of $W(n,m)$ is given by

$$p\left[W(n,m)\right] = \frac{1}{\pi}\exp\left[-|W(n,m)|^2\right]. \qquad (11.53)$$

The first assumption requires that the overlap between successive time frames would be small (less than 50%) [31]. The second assumption implies that the speech spectral coefficients $\{X(n,m)\}$ are conditionally zero-mean statistically independent random variables given their variances $\{\phi_x(n,m)\}$, satisfying

$$p\left[X(n,m)\mid \phi_x(n,m)\right] = \frac{1}{\pi\,\phi_x(n,m)}\exp\left[-\frac{|X(n,m)|^2}{\phi_x(n,m)}\right]. \qquad (11.54)$$

Note that successive spectral coefficients are generally correlated, since the random processes $\{X(n,m)\mid n = 0,1,\ldots\}$ and $\{\phi_x(n,m)\mid n = 0,1,\ldots\}$ are not independent. However, given $\phi_x(n,m)$, $X(n,m)$ is statistically independent of $X(n',m')$ for all $n \neq n'$ and $m \neq m'$. Hence, the time-correlation between successive spectral coefficients of the speech signal is taken into account, while still considering the scalar estimation problem formulated in (11.9).

11.5 Signal Estimation

In this section, we derive estimators for $X(n,m)$ using various fidelity criteria, assuming that $\hat{\phi}_x(n,m)$ and $\hat{\phi}_v(n,m)$ are given. Fidelity criteria that are of particular interest for speech enhancement applications are minimum mean-squared error (MMSE) [85], MMSE of the spectral amplitude (MMSE-SA) [44], and MMSE of the log-spectral amplitude (MMSE-LSA) [27], [45]. The MMSE estimator is derived by substituting into (11.11) the function $g[X(n,m)] = X(n,m)$. The MMSE-SA estimator is obtained by using $g[X(n,m)] = |X(n,m)|$, and the MMSE-LSA estimator is obtained by using $g[X(n,m)] = \log|X(n,m)|$.

11.5.1 MMSE Spectral Estimation

Let $H_{\mathrm{MSE}}(n,m)$ denote a gain function that satisfies

$$E\left[X(n,m) \mid \phi_x(n,m), \phi_v(n,m), Y(n,m)\right] = H_{\mathrm{MSE}}(n,m)Y(n,m). \quad (11.55)$$

Then, by substituting $g[X(n,m)] = X(n,m)$ into (11.11) we have

$$Z(n,m) = H_{\mathrm{MSE}}(n,m)\, Y(n,m). \quad (11.56)$$

The specific expression for $H_{\mathrm{MSE}}(n,m)$ depends on the statistical model:

$$H_{\mathrm{MSE}}(n,m) = \frac{1}{Y(n,m)} \int X(n,m)\, p\left[X(n,m) \mid \phi_x(n,m), \phi_v(n,m)\right] dX,$$

where

$$p\left[X(n,m) \mid \phi_x(n,m), \phi_v(n,m)\right] = \\ \frac{p\left[Y(n,m) \mid X(n,m), \phi_v(n,m)\right] p\left[X(n,m) \mid \phi_x(n,m)\right]}{p\left[Y(n,m) \mid \phi_x(n,m), \phi_v(n,m)\right]}.$$

For a Gaussian model, the gain function depends only on the input SNR [85]:

$$H_{\mathrm{MSE}}(n,m) = \frac{\mathrm{iSNR}(n,m)}{1 + \mathrm{iSNR}(n,m)}, \quad (11.57)$$

and the MMSE estimator for $X(n,m)$ reduces to

$$Z(n,m) = \frac{\widehat{\mathrm{iSNR}}(n,m)}{1 + \widehat{\mathrm{iSNR}}(n,m)}\, Y(n,m), \quad (11.58)$$

where $\widehat{\mathrm{iSNR}}(n,m) = \hat{\phi}_x(n,m)/\hat{\phi}_v(n,m)$ is an estimate of $\mathrm{iSNR}(n,m)$.

11.5.2 MMSE Spectral Amplitude Estimation

Estimators which minimize the MSE of the spectral amplitude, or log-spectral amplitude, have been found advantageous to MMSE spectral estimators in speech enhancement applications [44], [45], [100]. An MMSE-SA estimator is obtained by substituting $g[X(n,m)] = |X(n,m)| = A_x(n,m)$ into (11.11), and combining the resulting amplitude estimate with the phase of the noisy spectral coefficient $Y(n,m)$.

Let

$$\gamma(n,m) = \frac{|Y(n,m)|^2}{\phi_v(n,m)} \quad (11.59)$$

denote an instantaneous input SNR. Note that

$$E\left[\gamma(n,m)\right] = \frac{\phi_y(n,m)}{\phi_v(n,m)} = \frac{\phi_x(n,m) + \phi_v(n,m)}{\phi_v(n,m)} = \mathrm{iSNR}(n,m) + 1.$$

The spectral gain that satisfies

164   11 Spectral Enhancement Methods

$$E\left[A_x(n,m) \mid \phi_x(n,m), \phi_v(n,m), Y(n,m)\right] = H_{\mathrm{SA}}(n,m)A_y(n,m) \quad (11.60)$$

is given by [44]

$$
\begin{aligned}
H_{\mathrm{SA}}(n,m) &= \frac{1}{A_y(n,m)} \int A_x(n,m)\, p\left[X(n,m) \mid \phi_x(n,m), \phi_v(n,m)\right] dX \\
&= \frac{\sqrt{\pi}}{2} \frac{\sqrt{\vartheta(n,m)}}{\gamma(n,m)} \exp\left[-\frac{\vartheta(n,m)}{2}\right] \\
&\quad \times \left\{ [1+\vartheta(n,m)]I_0\left[\frac{\vartheta(n,m)}{2}\right] + \vartheta(n,m)\, I_1\left[\frac{\vartheta(n,m)}{2}\right] \right\},
\end{aligned}
$$

$$(11.61)$$

where

$$\vartheta(n,m) = \frac{\gamma(n,m)\,\mathrm{iSNR}(n,m)}{1+\mathrm{iSNR}(n,m)}, \quad (11.62)$$

and $I_0(\cdot)$ and $I_1(\cdot)$ denote the modified Bessel functions of zero and first order, respectively. Combining the resulting amplitude estimate with the phase of the noisy spectral coefficient yields

$$
\begin{aligned}
Z(n,m) &= \frac{\sqrt{\pi}}{2} \frac{\sqrt{\hat\vartheta(n,m)}}{\hat\gamma(n,m)} \exp\left[-\frac{\hat\vartheta(n,m)}{2}\right] \times \\
&\quad \left\{ [1+\hat\vartheta(n,m)]I_0\left[\frac{\hat\vartheta(n,m)}{2}\right] + \hat\vartheta(n,m)\, I_1\left[\frac{\hat\vartheta(n,m)}{2}\right] \right\} Y(n,m),
\end{aligned}
$$

$$(11.63)$$

where

$$\hat\gamma(n,m) = \frac{|Y(n,m)|^2}{\hat\phi_v(n,m)}, \quad \hat\vartheta(n,m) = \frac{\hat\gamma(n,m)\,\widehat{\mathrm{iSNR}}(n,m)}{1+\widehat{\mathrm{iSNR}}(n,m)}.$$

Figure 11.1 displays parametric gain curves describing $H_{\mathrm{SA}}(n,m)$ for several values of $\gamma(n,m)$. For a fixed value of the instantaneous input SNR, the LSA gain is a monotonically increasing function of $\mathrm{iSNR}(n,m)$. However, for a fixed value of $\mathrm{iSNR}(n,m)$, the LSA gain is a monotonically decreasing function of $\gamma(n,m)$.

11.5.3 MMSE Log-Spectral Amplitude Estimation

An MMSE-LSA estimator is obtained by substituting $g[X(n,m)] = \log[A_x(n,m)]$ into (11.11), and combining the resulting amplitude estimate with the phase of the noisy spectral coefficient. The spectral gain that satisfies

11.5 Signal Estimation   165

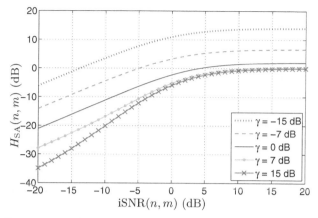

Fig. 11.1. Parametric gain curves describing the MMSE spectral amplitude gain function $H_{\text{SA}}(n,m)$ for a Gaussian model, obtained by (11.61).

$$E\left\{\log[A_x(n,m)] \mid \phi_x(n,m), \phi_v(n,m), Y(n,m)\right\} = H_{\text{LSA}}(n,m)A_y(n,m) \tag{11.64}$$

is given by [45]

$$H_{\text{LSA}}(n,m) = \frac{1}{A_y(n,m)} \int \log[A_x(n,m)] \, p\left[X(n,m) \mid \phi_x(n,m), \phi_v(n,m)\right] dX$$

$$= \frac{\text{iSNR}(n,m)}{1+\text{iSNR}(n,m)} \exp\left(\frac{1}{2}\int_{\vartheta(n,m)}^{\infty} \frac{e^{-x}}{x} dx\right). \tag{11.65}$$

The integral in (11.65) is the well known exponential integral of $\vartheta(n,m)$ and it can be numerically evaluated, e.g., using the *expint* function in MATLAB. Alternatively, it may be evaluated by using the following computationally efficient approximation, which was developed by Martin et al. [91]

$$\text{expint}(\vartheta) = \int_\vartheta^\infty \frac{e^{-x}}{x} dx \approx \begin{cases} -2.31 \log_{10}(\vartheta) - 0.6, & \text{for } \vartheta < 0.1 \\ -1.544 \log_{10}(\vartheta) + 0.166, & \text{for } 0.1 \leq \vartheta \leq 1 \\ 10^{-0.52\,\vartheta - 0.26}, & \text{for } \vartheta > 1 \end{cases} \tag{11.66}$$

Combining the resulting amplitude estimate with the phase of the noisy spectral coefficient yields

$$Z(n,m) = \frac{\widehat{\text{iSNR}}(n,m)}{1+\widehat{\text{iSNR}}(n,m)} \exp\left(\frac{1}{2}\int_{\hat{\vartheta}(n,m)}^{\infty} \frac{e^{-x}}{x} dx\right) Y(n,m). \tag{11.67}$$

Figure 11.2 displays parametric gain curves describing $H_{\text{LSA}}(n,m)$ for several values of $\gamma(n,m)$. For a fixed value of the instantaneous input SNR, the LSA gain is a monotonically increasing function of $\text{iSNR}(n,m)$. Similar to $H_{\text{SA}}(n,m)$, for a fixed value of $\text{iSNR}(n,m)$, the LSA gain is a monotonically

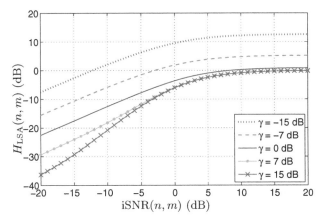

Fig. 11.2. Parametric gain curves describing the MMSE log-spectral amplitude gain function $H_{\text{LSA}}(n,m)$ for a Gaussian model, obtained by (11.65).

decreasing function of $\gamma(n,m)$. The behaviors of $H_{\text{SA}}(n,m)$ and $H_{\text{LSA}}(n,m)$ are related to the useful mechanism that counters the musical noise phenomenon [22]. Local bursts of the instantaneous input SNR, during noise-only frames, are "pulled down" to the average noise level, thus avoiding local buildup of noise whenever it exceeds its average characteristics. As a result, the MMSE-SA and MMSE-LSA estimators generally produce lower levels of residual musical noise, when compared with the MMSE spectral estimators.

11.6 Spectral Variance Model

In this section, we present a novel modeling approach for the speech spectral component variances $\phi_x(n,m) = E\left[|X(n,m)|^2\right]$. This approach is based on generalized autoregressive conditional heteroscedasticity (GARCH) modeling, which is widely-used for modeling the volatility of financial time-series such as exchange rates and stock returns [20], [43]. Similar to financial time-series, speech signals in the STFT domain are characterized by heavy tailed distributions and volatility clustering [30]. GARCH models enable parametrization of the time variation of speech spectral variances, and derivation of mathematically tractable estimates, while taking into consideration the heavy-tailed distribution and variation clustering.

11.6.1 GARCH Model

Let $\{d(k)\}$ denote a real-valued discrete-time stochastic process and let $\psi(k)$ denote the information set available at time k [e.g., $\{d(k)\}$ may represent a sequence of observations, and $\psi(k)$ may include the observed data through time k]. Then, the innovation (prediction error) $\varepsilon(k)$ at time k in the MMSE

11.6 Spectral Variance Model   167

sense is obtained by subtracting from $d(k)$ its conditional expectation given the information through time $k - 1$,

$$\varepsilon(k) = d(k) - E\left[d(k) \mid \psi(k-1)\right]. \tag{11.68}$$

The conditional variance (volatility) of $d(k)$ given the information through time $k - 1$ is by definition the conditional expectation of $\varepsilon^2(k)$,

$$\sigma^2(k) = \text{var}\left[d(k) \mid \psi(k-1)\right] = E\left[\varepsilon^2(k) \mid \psi(k-1)\right]. \tag{11.69}$$

Autoregressive conditional heteroscedasticity (ARCH) models, introduced by Engle [42] and generalized by Bollerslev [19], are widely-used in various financial applications such as risk management, option pricing, foreign exchange, and the term structure of interest rates [20]. They explicitly parameterize the time-varying volatility in terms of past conditional variances and past squared innovations (prediction errors), while taking into account excess kurtosis (i.e., heavy tail behavior) and volatility clustering, two important characteristics of financial time-series. ARCH and GARCH models explicitly recognize the difference between the unconditional variance $E\left\{|d(k) - E[d(k)]|^2\right\}$ and the conditional variance $\sigma^2(k)$, allowing the latter to change over time. The fundamental characteristic of these models is that magnitudes of recent innovations provide information about future volatility.

Let $\{w(k)\}$ be a zero-mean unit-variance white noise process with some specified probability distribution. Then a GARCH model of order (p, q), denoted by $\varepsilon(k) \sim \text{GARCH}(p, q)$, has the following general form

$$\varepsilon(k) = \sigma(k)\, w(k), \tag{11.70}$$
$$\sigma^2(k) = f\left[\sigma^2(k-1), \ldots, \sigma^2(k-p), \varepsilon^2(k-1) \ldots, \varepsilon^2(k-q)\right], \tag{11.71}$$

where $\sigma(k)$ is the conditional standard deviation given by the square root of (11.71). That is, the conditional variance $\sigma^2(k)$ is determined by the values of p past conditional variances and q past squared innovations, and the innovation $\varepsilon(k)$ is generated by scaling a white noise sample with the conditional standard deviation. The ARCH(q) model is a special case of the GARCH(p, q) model with $p = 0$.

The most widely-used GARCH model specifies a linear function f in (11.71) as follows,

$$\sigma^2(k) = \kappa + \sum_{i=1}^{q} \alpha_i\, \varepsilon^2(k - i) + \sum_{j=1}^{p} \beta_j\, \sigma^2(k - j), \tag{11.72}$$

where the values of the parameters are constrained by

$$\kappa > 0,\, \alpha_i \geq 0,\, \beta_j \geq 0, \quad i = 1, \ldots, q,\, j = 1, \ldots, p,$$
$$\sum_{i=1}^{q} \alpha_i + \sum_{j=1}^{p} \beta_j < 1.$$

168   11 Spectral Enhancement Methods

The first three constraints are sufficient to ensure that the conditional variances $\{\sigma^2(k)\}$ are strictly positive. The forth constraint is a covariance stationarity constraint, which is necessary and sufficient for the existence of a finite unconditional variance of the innovations process [19].

Many financial time-series such as exchange rates and stock returns exhibit volatility clustering phenomenon, i.e., large changes tend to follow large changes of either sign and small changes tend to follow small changes. Equation (11.72) captures the volatility clustering phenomenon, since large innovations of either sign increase the variance forecasts for several samples. This in return increases the likelihood of large innovations in the succeeding samples, which allows the large innovations to persist. The degree of persistence is determined by the lag lengths p and q, as well as the magnitudes of the coefficients $\{\alpha_i\}$ and $\{\beta_j\}$. Furthermore, the innovations of financial time-series are typically distributed with heavier tails than a Gaussian distribution. Bollerslev [19] showed that GARCH models are appropriate for heavy-tailed distributions.

11.6.2 Modeling Speech Spectral Variance

The variances of the speech coefficients are hidden from direct observation, in the sense that even under perfect conditions of zero noise $[V(n, m) = 0$ $\forall(n, m)]$, the values of $\{\phi_x(n, m)\}$ are not directly observable. Therefore, our approach is to assume that $\{\phi_x(n, m)\}$ themselves are random variables, and to introduce *conditional* variances which are estimated from the available information (e.g., the clean spectral coefficients through time-frame $n - 1$, or the noisy spectral coefficients through time-frame n) [30]. Let $\mathcal{X}_0^r = \{X(n, m) \,|\, n = 0, \ldots, r, \ m = 0, \ldots, M - 1\}$ represent the set of clean speech spectral coefficients up to frame r, and let $\phi_{x|r}(n, m) \triangleq E\left\{A_x^2(n, m) \,|\, \mathcal{X}_0^r\right\}$ denote the *conditional* variance of $X(n, m)$ given the clean spectral coefficients up to frame r. Then, our statistical model for the speech spectral variance relies on the following assumption [30]: the conditional variance $\phi_{x|n-1}(n, m)$, referred to as the *one-frame-ahead conditional variance*, is a random process which evolves as a GARCH(1, 1) process:

$$\phi_{x|n-1}(n, m) = \phi_{\min} + \mu \, A_x^2(n - 1, m) + \delta \left[\phi_{x|n-2}(n - 1, m) - \phi_{\min}\right], \quad (11.73)$$

where

$$\phi_{\min} > 0, \quad \mu \geq 0, \quad \delta \geq 0, \quad \mu + \delta < 1, \quad (11.74)$$

are the standard constraints imposed on the parameters of the GARCH model [20]. The parameters μ and δ are, respectively, the moving average and autoregressive parameters of the GARCH(1,1) model, and $\phi_{\min}$ is a lower bound on the variance of $X(n, m)$.

11.6.3 Model Estimation

The maximum-likelihood (ML) estimation approach is commonly used for estimating the parameters of a GARCH model [43]. In this section we derive the ML function of the model parameters, by using the spectral coefficients of the clean speech signal on some interval $n \in [0, N-1]$. For simplicity, we assume that the parameters are constant during the above interval and are independent of the frequency-bin index m. In practice, the speech signal can be divided into short time segments and split in frequency into narrow subbands, such that the parameters can be assumed to be constant in each time-frequency region. Furthermore, we generally do not have a direct access to the clean spectral coefficients. However, the expectation-maximization (EM) algorithm [36], [93] can be utilized for solving this problem by iteratively estimating the clean spectral coefficients and the model parameters from the noisy measurements.

Let $\mathcal{X}_0^{N-1}$ denote the set of clean speech spectral coefficients employed for the model estimation, let $\mathcal{H}_1$ denote the set of time-frequency bins where the signal is present, and let $\boldsymbol{\lambda} = \begin{bmatrix} \mu & \delta & \phi_{\min} \end{bmatrix}^T$ denote the vector of unknown parameters. Then, the conditional variance $\phi_{x|n-1}(n, m)$ can recursively be calculated from past spectral coefficients $\mathcal{X}_0^{n-1}$ by using (11.73) and the parameter vector $\boldsymbol{\lambda}$. Hence, for a Gaussian model, the logarithm of the conditional density of $X(n, m)$ given the clean spectral coefficients up to frame $n-1$ can be expressed as

$$\log p\left[X(n, m) \mid \mathcal{X}_0^{n-1} ; \boldsymbol{\lambda}\right] = -\frac{A_x^2(n, m)}{\phi_{x|n-1}(n, m)} - \log \phi_{x|n-1}(n, m) - \log \pi,$$

$$(11.75)$$

where $(n, m) \in \mathcal{H}_1$. It is convenient to regard the speech spectral coefficients in the first frame ($n = 0$) as deterministic, with the values of $\phi_{x|-1}(0, m)$ in the first frame initialized to their minimal value $\phi_{\min}$, and maximize the log-likelihood when conditioned on the first frame (for sufficiently large sample size, the spectral coefficients of the first frame make a negligible contribution to the total likelihood). The log-likelihood conditional on the spectral coefficients of the first frame is given by

$$\mathcal{L}(\boldsymbol{\lambda}) = \sum_{(n,m)\in\mathcal{H}_1\cap n\in[1,N-1]} \log p\left[X(n, m) \mid \mathcal{X}_0^{n-1} ; \boldsymbol{\lambda}\right]. \qquad (11.76)$$

Substituting (11.75) into (11.76) and imposing the constraints in (11.74) on the estimated parameters, the ML estimates of the model parameters can be obtained by solving a constrained minimization problem:

$$\underset{\hat{\phi}_{\min}, \hat{\mu}, \hat{\delta}}{\text{minimize}} \sum_{(n,m)\in\mathcal{H}_1\cap n\in[1,N-1]} \left[\frac{A_x^2(n, m)}{\phi_{x|n-1}(n, m)} + \log \phi_{x|n-1}(n, m)\right] \qquad (11.77)$$

subject to the constraints

170   11 Spectral Enhancement Methods

$$\hat{\phi}_{\min} > 0, \; \hat{\mu} \geq 0, \; \hat{\delta} \geq 0, \; \hat{\mu} + \hat{\delta} < 1. \tag{11.78}$$

For given numerical values of the parameters, the sequences of conditional variances $\{\phi_{x|n-1}(n, m)\}$ can be calculated from (11.73) and used to evaluate the series in (11.77). The result can then be minimized numerically by using the Berndt, Hall, Hall, and Hausman [16] algorithm as in Bollerslev [19]. Alternatively, the function *fmincon* of the Optimization Toolbox in MATLAB can be used to find the minimum of the constrained nonlinear function of the model parameters, similar to its use within the function *garchfit* of the GARCH Toolbox.

11.7 Spectral Variance Estimation

In this section, we assume that the model parameters μ, δ, and $\phi_{\min}$ are known, and derive a recursive estimator for the speech spectral variance. The speech variance estimation approach is closely related to the variance estimation approach introduced in [30], [31]. We start with an estimate $\hat{\phi}_{x|n-1}(n, m)$ that relies on the noisy observations up to time-frame $n-1$, and "update" the variance by using the additional information $Y(n, m)$. Then, the variance is "propagated" ahead in time, following the rational of Kalman filtering, to obtain a conditional variance estimate at time-frame $n+1$ from the information available at time-frame n.

Assuming an estimate $\hat{\phi}_{x|n-1}(n, m)$ for the one-frame-ahead conditional variance of $X(n, m)$ is available, an estimate for $\phi_{x|n}(n, m)$ can be obtained by calculating its conditional mean given $Y(n, m)$ and $\hat{\phi}_{x|n-1}(n, m)$. By definition, $\phi_{x|n}(n, m) = A_x^2(n, m)$. Hence,

$$\hat{\phi}_{x|n}(n, m) = E\left[A_x^2(n, m) \; \middle| \; \hat{\phi}_{x|n-1}(n, m), Y(n, m) \right]. \tag{11.79}$$

We can write

$$E\left[A_x^2(n, m) \; \middle| \; \hat{\phi}_{x|n-1}(n, m), Y(n, m) \right] = H_{\mathrm{SP}}(n, m) A_y^2(n, m), \tag{11.80}$$

where $H_{\mathrm{SP}}(n, m)$ represents the MMSE gain function in the spectral power domain [32]. The specific expression for $H_{\mathrm{SP}}(n, m)$ depends on the particular statistical model. For a Gaussian model, the spectral power gain function is given by

$$
\begin{aligned}
H_{\mathrm{SP}}(n, m) &= \frac{1}{A_y^2(n, m)} \int A_x^2(n, m) \, p\left[X(n, m) \mid \phi_x(n, m), \phi_v(n, m) \right] dX \\
&= \frac{\mathrm{iSNR}(n, m)}{1 + \mathrm{iSNR}(n, m)} \left[\frac{1}{\gamma(n, m)} + \frac{\mathrm{iSNR}(n, m)}{1 + \mathrm{iSNR}(n, m)} \right]. \tag{11.81}
\end{aligned}
$$

Substituting (11.80) into (11.79), we obtain an estimate for $\phi_{x|n}(n, m)$ given by

$$\hat{\phi}_{x|n}(n,m) = \frac{\widehat{\text{iSNR}}_{n-1}(n,m)}{1+\widehat{\text{iSNR}}_{n-1}(n,m)}\left[\frac{1}{\gamma(n,m)} + \frac{\widehat{\text{iSNR}}_{n-1}(n,m)}{1+\widehat{\text{iSNR}}_{n-1}(n,m)}\right]A_y^2(n,m)$$

$$= \frac{\hat{\phi}_{x|n-1}(n,m)\phi_v(n,m)}{\phi_v(n,m)+\hat{\phi}_{x|n-1}(n,m)} + \left[\frac{\hat{\phi}_{x|n-1}(n,m)A_y(n,m)}{\phi_v(n,m)+\hat{\phi}_{x|n-1}(n,m)}\right]^2.$$

$$(11.82)$$

Equation (11.82) is the update step of the recursive estimation, since we start with an estimate $\hat{\phi}_{x|n-1}(n,m)$ that relies on the noisy observations up to time-frame $n-1$, and then update the estimate by using the additional information $Y(n,m)$.

To formulate the propagation step, we assume that we are given at time-frame $n-1$ an estimate $\hat{\phi}_{x|n-2}(n-1,m)$ for the conditional variance of $X(n-1,m)$, which has been obtained from the noisy measurements up to frame $n-2$. Then a recursive MMSE estimate for $\hat{\phi}_{x|n-1}(n,m)$ can be obtained by calculating its conditional mean given $\hat{\phi}_{x|n-2}(n-1,m)$ and $Y(n-1,m)$:

$$\hat{\phi}_{x|n-1}(n,m) = E\left[\phi_{x|n-1}(n,m)\,\Big|\,\hat{\phi}_{x|n-2}(n-1,m),Y(n-1,m)\right]. \quad (11.83)$$

Substituting (11.73) into (11.83), we have

$$\hat{\phi}_{x|n-1}(n,m) = \phi_{\min} + \mu\,E\left[A_x^2(n-1,m)\,\Big|\,\hat{\phi}_{x|n-2}(n-1,m),Y(n-1,m)\right]$$
$$+ \delta\left[\hat{\phi}_{x|n-2}(n-1,m) - \phi_{\min}\right]. \quad (11.84)$$

Equation (11.79) implies that

$$E\left[A_x^2(n-1,m)\,\Big|\,\hat{\phi}_{x|n-2}(n-1,m),Y(n-1,m)\right] = \hat{\phi}_{x|n-1}(n-1,m).$$

Substituting this into (11.84), we obtain

$$\hat{\phi}_{x|n-1}(n,m) = \phi_{\min} + \mu\,\hat{\phi}_{x|n-1}(n-1,m) + \delta\left[\hat{\phi}_{x|n-2}(n-1,m) - \phi_{\min}\right].$$

$$(11.85)$$

Equation (11.85) is called the propagation step, since the conditional variance estimates are propagated ahead in time to obtain a conditional variance estimate at time-frame n from the information available at time-frame $n-1$. The propagation and update steps are iterated as new data arrive, following the rational of Kalman filtering. The algorithm is initialized at the first time-frame, say $n=0$, with $\hat{\phi}_{x|-1}(0,m) = \phi_{\min}$ for all the frequency bins, $m = 0,\ldots,M-1$. Then, for $n = 0,1,\ldots$, the estimate $\hat{\phi}_{x|n}(n,m)$ is calculated by using the update step (11.82) and $\hat{\phi}_{x|n}(n+1,m)$ is subsequently calculated by using the propagation step (11.85).

172   11 Spectral Enhancement Methods

11.7.1 Relation to Decision-Directed Estimation

The recursive spectral variance estimator $\hat{\phi}_{x|n}(n, m)$ is closely related to the decision-directed estimator of Ephraim and Malah [44]. The decision-directed estimator is given by

$$\hat{\phi}_x^{\mathrm{DD}}(n, m) = \alpha \hat{A}_x^2(n-1, m) + (1-\alpha) \max \left\{ A_y^2(n, m) - \phi_v(n, m), \beta\phi_v(n, m) \right\},$$
(11.86)

where α $(0 \le \alpha \le 1)$ is a weighting factor that controls the tradeoff between noise reduction and transient distortion introduced into the signal. The parameter β is a lower bound on the estimated input SNR to further reduce the level of residual musical noise [22]. The decision-directed estimator is particularly useful when combined with the MMSE spectral, or log-spectral, magnitude estimators [22], [44], [45]. It results in perceptually colorless residual noise, but is heuristically motivated and its theoretical performance is unknown due to its highly nonlinear nature. Furthermore, the parameters of the decision-directed estimator have to be determined by simulations and subjective listening tests for each particular setup of time-frequency transformation and speech enhancement algorithm. Since the decision-directed approach is not supported by a statistical model, the parameters are not adapted to the speech components, but are set to specific values in advance.

The update step (11.82) of the recursive estimator can be written as

$$\hat{\phi}_{x|n}(n, m) = \alpha(n, m)\, \hat{\phi}_{x|n-1}(n, m) + [1 - \alpha(n, m)] \left[A_y^2(n, m) - \phi_v(n, m) \right],$$
(11.87)

where $\alpha(n, m)$ is given by

$$\alpha(n, m) = 1 - \frac{\hat{\phi}_{x|n-1}(n, m)}{\left[\phi_v(n, m) + \hat{\phi}_{x|n-1}(n, m) \right]^2}.$$
(11.88)

Substituting (11.84) into (11.87) with $\mu \equiv 1$ we have

$$\hat{\phi}_{x|n}(n, m) = \alpha(n, m)\, E\left[A_x^2(n-1, m) \,\middle|\, \hat{\phi}_{x|n-2}(n-1, m), Y(n-1, m) \right]$$
$$+ [1 - \alpha(n, m)] \left[A_y^2(n, m) - \phi_v(n, m) \right] + \alpha(n, m)\phi_{\min}. (11.89)$$

The expression (11.89) is an alternative practical form of the decision-directed estimator, with α replaced by $\alpha(n, m)$, $A_x^2(n-1, m)$ replaced by $E\left[A_x^2(n-1, m) \,\middle|\, \hat{\phi}_{x|n-2}(n-1, m), Y(n-1, m) \right]$, and instead of the parameter β, which represents a lower bound on the estimated input SNR $\hat{\phi}_x^{\mathrm{DD}}(n, m)/\phi_v(n, m)$, we have a parameter $\phi_{\min}$, which is a lower bound on $\hat{\phi}_{x|n-1}(n, m)$. Accordingly, a special case of the recursive estimator with $\mu \equiv 1$ degenerates to a "decision-directed" estimator with a *time-varying frequency-dependent* weighting factor $\alpha(n, m)$.

It is interesting to note that the weighting factor $\alpha(n, m)$, given by (11.88), is monotonically decreasing as a function of the one-frame-ahead conditional

11.7 Spectral Variance Estimation 173

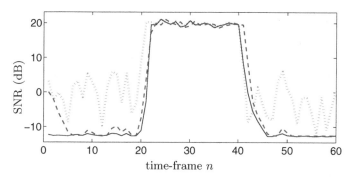

Fig. 11.3. SNR in successive time-frames: instantaneous input SNR, $\gamma(n,m)$ (dotted line); decision-directed estimate of the input SNR, $\widehat{\text{iSNR}}^{\text{DD}}(n,m)$ (dashed line); and recursive estimate of the input SNR, $\widehat{\text{iSNR}}_n(n,m)$ (solid line).

input SNR, $\widehat{\text{iSNR}}_{n-1}(n,m) = \hat{\phi}_{x|n-1}(n,m)/\phi_v(n,m)$. A decision-directed estimator with a larger weighting factor is indeed preferable during speech absence (to reduce musical noise phenomena), while a smaller weighting factor is more advantageous during speech presence (to reduce signal distortion) [22]. The above special case of the recursive estimator conforms to such a desirable behavior. Moreover, the general form of the recursive estimator provides an additional degree of freedom for adjusting the value of μ in (11.85) to the degree of spectral nonstationarity. This may produce even further improvement in the performance.

The different behaviors of the recursive estimator $\hat{\phi}_{x|n}(n,m)$ (11.82) and the decision-directed estimator $\hat{\phi}_x^{\text{DD}}(n,m)$ (11.86) are illustrated in the example of Fig. 11.3. The analyzed signal contains only white Gaussian noise during the first and last 20 frames, and in between it contains an additional sinusoidal component at the displayed frequency with 0 dB SNR. The signal is transformed into the STFT domain using half overlapping Hamming windows. The SNR estimates, $\widehat{\text{iSNR}}_n(n,m) = \hat{\phi}_{x|n}(n,m)/\phi_v(n,m)$ and $\widehat{\text{iSNR}}^{\text{DD}}(n,m) = \hat{\phi}_x^{\text{DD}}(n,m)/\phi_v(n,m)$, are obtained by using the parameters $\beta = -25$ dB, $\alpha = 0.98$, $\mu = 0.9$. The signal estimate $Z(n,m)$ is recursively obtained by applying $H_{\text{LSA}}(n,m)$ to the noisy spectral measurements [see (11.67)].

Figure 11.3 shows that when the instantaneous SNR $\gamma(n,m)$ is sufficiently low, the recursive input SNR estimate is smoother than the decision-directed estimate, which helps reducing the level of musical noise. When $\gamma(n,m)$ increases, the response of $\widehat{\text{iSNR}}_n(n,m)$ is initially slower than $\widehat{\text{iSNR}}^{\text{DD}}(n,m)$, but it then builds up faster to the instantaneous SNR. When $\gamma(n,m)$ is sufficiently high, $\widehat{\text{iSNR}}^{\text{DD}}(n,m)$ follows the instantaneous SNR with a delay of 1 frame, whereas $\widehat{\text{iSNR}}_n(n,m)$ follows the instantaneous SNR without delay.

174   11 Spectral Enhancement Methods

When $\gamma(n, m)$ decreases, the response of $\widehat{\text{iSNR}}_n(n, m)$ is immediate, while that of $\widehat{\text{iSNR}}^{\text{DD}}(n, m)$ is delayed by 1 frame. As a consequence, when compared with the decision-directed estimator, the recursive estimator produces a lower level of musical noise while not increasing the audible distortion in the enhanced signal [31].

11.8 Summary of Spectral Enhancement Algorithm

In this section, we present an example of a speech enhancement algorithm, which is based on an MMSE log-spectral amplitude estimation under a Gaussian model, improved minima-controlled recursive averaging (IMCRA) noise estimation [28], and recursive estimation of the input SNR. The performance of the algorithm is demonstrated on speech signals degraded by various additive noise types.

The implementation of the speech enhancement algorithm is summarized in Table 11.1. For each time-frame n we recursively estimate the STFT coefficients of the clean speech $\{X(n, m) \,|\, m = 0, \ldots, M - 1\}$ from the noisy STFT coefficients $\{Y(n, m) \,|\, k = 0, \ldots, M - 1\}$, where M is the length of the analysis window. We typically use a Hamming window of 32 ms length and a framing step of 8 ms (i.e., $M = 512$ and $L = 128$ for a sampling rate of 16 kHz). In the first frame ($n = 0$) we compute $\{Y(0, m) \,|\, m = 0, \ldots, M - 1\}$ by applying the discrete Fourier transform to a short-time section of the noisy data

$$\mathbf{y}_\psi(0) = [y(0)\psi(0) \quad y(1)\psi(1) \quad \cdots \quad y(M - 1)\psi(M - 1)]^T,$$

where $\psi(n)$ is the analysis window. In the following frames ($n > 0$), the section of noisy data is updated with L additional samples

$$\mathbf{y}_\psi(n) = [y(nL)\psi(0) \quad y(1+nL)\psi(1) \quad \cdots \quad y(M-1+nL)\psi(M-1)]^T \quad (11.90)$$

and subsequently $\{Y(n, m) \,|\, m = 0, \ldots, M - 1\}$ is computed by applying the discrete Fourier transform to $\mathbf{y}_\psi(n)$. Since the speech signal $x(k)$ is assumed real, once we have spectral coefficients $\{Z(n, m) \,|\, m = 1, \ldots, (M - 1)/2\}$, the spectral coefficients for $(M - 1)/2 < m \leq M - 1$ are obtained by $Z(n, m) = Z^*(n, M - m)$. The DC component $Z(n, 0)$ is set to zero, and a sequence $\{z_n(k) \,|\, k = 0, \ldots, M - 1\}$ is obtained by applying the inverse discrete Fourier transform to $\{Z(n, m) \,|\, m = 0, \ldots, M - 1\}$:

$$z_n(k) = \frac{1}{M} \sum_{m=0}^{M-1} Z(n, m) \, e^{j\frac{2\pi}{M} km}. \quad (11.91)$$

Employing the weighted overlap-add method [35], we compute the following sequence

11.8 Summary of Spectral Enhancement Algorithm 175

Table 11.1. Summary of a spectral enhancement algorithm.

Initialization at the first time-frame $n = 0$:

$\hat{\phi}_v(0, m) = A_y^2(0, m)$; $\hat{\phi}_{x|-1}(-1, m) = \phi_{\min}$, for all m;

Read $M - L$ samples of the noisy signal $y(k)$.

For all time-frames $n = 0, 1, \ldots$

Read L new samples of the noisy signal, and compute $\mathbf{y}_\psi(n)$ using (11.90).

Compute $\{Y(n, m) \mid m = 0, \ldots, M - 1\}$ by applying the discrete Fourier transform to $\mathbf{y}_\psi(n)$.

For frequency-bins $m = 1, \ldots, (M - 1)/2$

Update the variance of the noise spectral coefficient $\hat{\phi}_v(n, m)$ using [28].

Compute $\hat{\phi}_{x|n-1}(n, m)$ using the propagation step of the recursive variance estimation (11.85).

Compute $\hat{\phi}_{x|n}(n, m)$ using the update step of the recursive variance estimation (11.82).

Compute $\widehat{\mathrm{iSNR}}(n, m) = \hat{\phi}_{x|n}(n, m) / \hat{\phi}_v(n, m)$, and compute $\hat{\vartheta}(n, m)$ using (11.62).

Compute the speech spectral estimate $Z(n, m)$ using (11.67).

Let $Z(n, 0) = 0$, and let $Z(n, m) = Z^*(n, M - m)$ for $m = (M + 1)/2, \ldots, M - 1$.

Apply the inverse discrete Fourier transform to $\{Z(n, m) \mid m = 0, \ldots, M - 1\}$, and compute L new samples of the enhanced speech signal using (11.91)–(11.93).

$$
o_n(k) = \begin{cases} o_{n-1}(k + L) + M \, \tilde{\psi}(k) \, z_n(k), & \text{for } 0 \le k \le M - L - 1 \\ M \, \tilde{\psi}(k) \, z_n(k), & \text{for } M - L \le k \le M - 1 \end{cases}, \quad (11.92)
$$

where $\tilde{\psi}(k)$ is the synthesis window. Then, according to (11.3), for each time-frame n, we obtain L additional samples of the enhanced speech signal:

$$
z(k + nL) = o_n(k), \quad k = 0, \ldots, L - 1. \tag{11.93}
$$

The synthesis window $\tilde{\psi}(k)$ should satisfy the completeness condition [124]

$$
\sum_n \tilde{\psi}(k - nL) \, \psi(k - nL) = \frac{1}{M} \quad \text{for all } k. \tag{11.94}
$$

Given analysis and synthesis windows that satisfy (11.94), any signal $x(k) \in \ell_2(\mathbb{Z})$ can be perfectly reconstructed from its STFT coefficients $X(n, m)$. However, for $L < M$ (over-redundant STFT representation) and for a given analysis window $\psi(k)$, there might be an infinite number of solutions to (11.94).

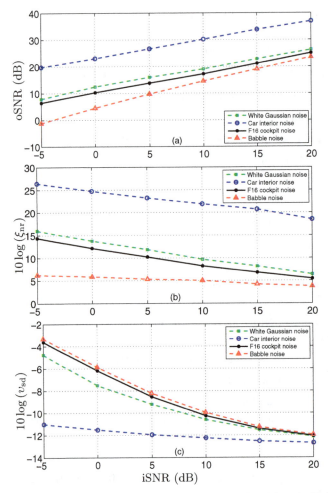

Fig. 11.4. Performance of the spectral enhancement algorithm in Table 11.1 in different input SNR conditions.

A reasonable choice of a synthesis window is the one with minimum energy [102], [124], given by

$$\tilde{\psi}(k) = \frac{\psi(k)}{M \sum_\ell \psi^2(k - \ell L)}. \tag{11.95}$$

11.9 Experimental Results

The speech signals used in our evaluation are taken from the TIMIT database [57]. They include 20 different utterances from 20 different speakers, half male and half female. The speech signals are sampled at 16 kHz and degraded by

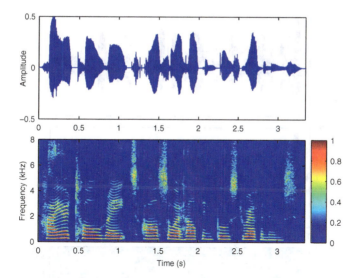

Fig. 11.5. Waveform and spectrogram of a clean speech signal: "Higher toll rates also are helping boost revenues."

various noise types from the Noisex92 database [118], which include white Gaussian noise, car interior noise, F16 cockpit noise, and babble noise. The input SNR is in the range $[-5, 20]$ dB. The average fullband output SNR, noise-reduction factor, and speech-distortion index obtained by the applying the spectral enhancement algorithm to the noisy signals are displayed in Fig. 11.4.

The results show that the fullband output SNR, oSNR(H), is larger than the fullband input SNR in all the tested conditions. However, the increase in SNR depends on the input SNR level and on the type of background noise. The increase in SNR is larger for lower input SNR levels and for noise characteristics which are different than the speech signal. The faster the noise spectrum varies in time, the less reliable is the noise spectrum estimator, and consequently the lower is the increase in SNR that can be achieved by the spectral enhancement algorithm. For car interior noise, most of the noise energy is concentrated in the lower frequencies, and its characteristics change slowly in time compared to the speech signal. Therefore, the output SNR and noise-reduction factor can be large, while keeping the speech-distortion index small. On the other hand, the characteristics of babble noise are similar to speech signals and the variations in time are faster, when compared to the other noise types. Therefore, the noise spectrum estimator is least reliable for babble noise and the speech enhancement performance is inferior to that achievable in the other noise environments. In particular, the noise-reduction factor is smallest for babble noise and largest for car interior noise, while the

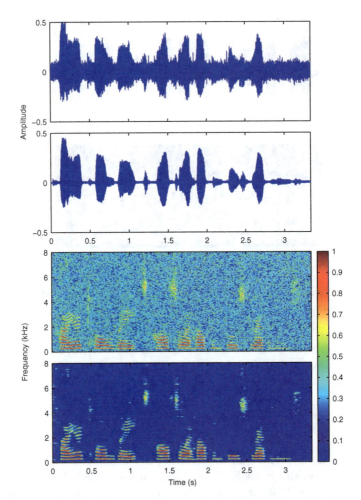

Fig. 11.6. Waveforms and spectrograms of the noisy and enhanced speech in white Gaussian noise environment with fullband iSNR = 5 dB (oSNR = 16.5 dB, ξ_{nr} = 12.1 dB, v_{sd} = −12.4 dB).

speech-distortion index is smallest for car interior noise and largest for babble noise.

A clean utterance from a female speaker is shown in Fig. 11.5. Figures 11.6–11.9 show its noisy versions in different environments (with input SNR of 5 dB) and the enhanced speech signals. The enhanced speech signals are obtained by using the spectral enhancement algorithm in Table 11.1. The fullband input SNR is the same for all the signals shown in Figs. 11.6–11.9, but the subband input SNR in the STFT domain, iSNR(n, m), varies significantly depending on the type of background noise. The noise spectrum is uniform in time and frequency for the white Gaussian noise. Hence, in time-frequency bins where

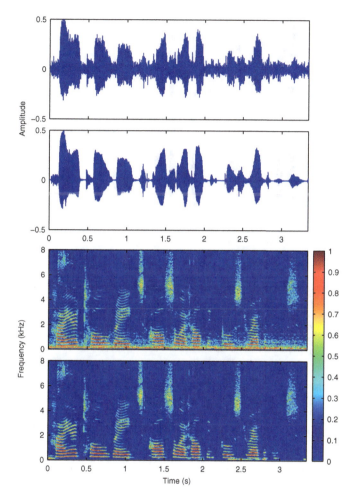

Fig. 11.7. Waveforms and spectrograms of the noisy and enhanced speech in car interior noise environment with fullband iSNR = 5 dB (oSNR = 16.3 dB, ξ_{nr} = 11.6 dB, v_{sd} = −17.5 dB).

the speech spectral variance $\phi_x(n,m)$ is high, the spectral gain $H(n,m)$ is closer to 1, in order to reduce the speech-distortion index. In time-frequency bins where the speech spectral variance $\phi_x(n,m)$ is low, the spectral gain $H(n,m)$ is closer to 0, in order to increase the noise-reduction factor. For car interior noise, one can apply a significant noise reduction in the lower frequencies and subtle noise reduction in the higher frequencies, since most of the noise energy is concentrated in the lower frequencies. Hence, the subband noise-reduction factors are large in the lower frequencies and small in the higher frequencies, while the subband speech-distortion indices are small in the higher frequencies and large in the lower frequencies. The fullband noise-

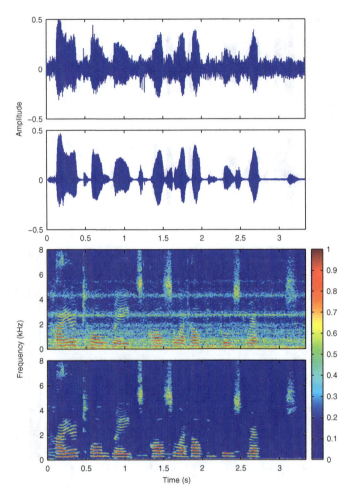

Fig. 11.8. Waveforms and spectrograms of the noisy and enhanced speech in F16 cockpit noise environment with fullband iSNR = 5 dB (oSNR = 14.7 dB, ξ_{nr} = 10.7 dB, v_{sd} = −12.7 dB).

reduction factor would generally be large, while keeping the fullband speech-distortion index low. For babble noise, the noise spectrum varies significantly in time and frequency. Hence, to restrict the fullband speech-distortion index to be smaller than a certain threshold, one has to restrict the spectral gain $H(n, m)$ to larger values, and thus restrict the fullband noise-reduction factor to smaller values. This generally yields higher residual noise levels than those achieved in other noise environments.

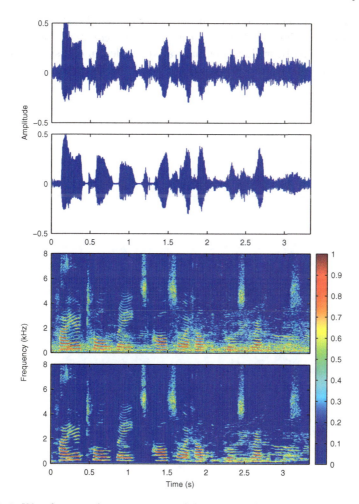

Fig. 11.9. Waveforms and spectrograms of the noisy and enhanced speech in babble noise environment with fullband iSNR = 5 dB (oSNR = 8.8 dB, ξ_{nr} = 4.4 dB, v_{sd} = −16.0 dB).

11.10 Summary

In this chapter, we formulated the noise reduction problem in the STFT domain and the derived MMSE, MMSE-SA and MMSE-LSA estimators for the speech spectral coefficients. The estimators rely on Gaussian statistical models for the speech and noise signals. The Gaussian model for the speech signal is integrated with a GARCH model for the spectral variances, thus enabling to take into account time-correlation between successive spectral components of the speech signal. We showed that the resulting recursive spectral variance estimator is closely related to the decision-directed estimator of Ephraim and

Malah. We described the beneficial behaviors of the MMSE-SA and MMSE-LSA estimators, which enable the mechanism that counters the musical noise phenomenon. Finally, we have demonstrated the design of a spectral enhancement algorithm and discussed its performance in different noisy environments.

12

A Practical Example: Multichannel Noise Reduction for Voice Communication in Spacesuits

Noise reduction can find numerous applications in various fields of speech processing. Since acoustic environments and technical objectives vary from application to application, it is impossible (at least impossible at the present time) to find a universally feasible approach. In the design of a microphone array system and the choice of a well-suited noise reduction algorithm, there is no simple rule to follow and experiences are still crucial. In this chapter, we use a practical example to explain the primary considerations that need to be kept in mind. We think that what can be interesting to the readers is not the finding of the best noise reduction algorithm for the specific problem under investigation but instead, it is the analysis of the acoustic challenges and the validation procedure of the suggested solutions that can be more useful and that can enrich the reader's fund of knowledge.

12.1 Problem Description

Collaboration and cooperation between the crewmembers in space and the mission control center on the earth are the lifeline of astronauts and space shuttles. Clear and reliable voice communications are essential to astronaut safety and the success of every NASA flight mission. But widely varying working conditions of a space shuttle and the special design of an astronaut's spacesuit form an extreme acoustic environment that imposes unique challenges for capturing and transmitting speech communications to and from a crewmember, as summarized in, but not limited to, the followings [2].

- Noise heard in the spacesuit has complicated characteristics in either the temporal, spectral, or spatial domain: generally non-stationary, inherently wideband, and possibly either directional or dispersive. In addition, during launch, entry descent, and landing, ambient noise penetrating from outside the spacesuit is at a very high level, while during on-orbit and Extravehicular Activity (EVA) operations suit-borne noise makes it difficult to achieve an adequate SNR for satisfactory voice communication.

J. Benesty et al., *Noise Reduction in Speech Processing*, Springer Topics in Signal Processing 2,
DOI 10.1007/978-3-642-00296-0_12, © Springer-Verlag Berlin Heidelberg 2009

184   12 Practical Example

- The helmet of a spacesuit is made of highly reflective materials, which create a very reverberant in-suit acoustic environment. Strong reverberation will severely distort the speech uttered by the suit subject and hence remarkably reduce its intelligibility. Moreover, strong reverberation leads to a more dispersive noise field, which makes beamforming less effective.
- Spacesuits can operate at several non-standard static pressures during EVA: specific static pressure levels include 30, 50, 55, 100, 101 kPa. These changes in static pressure level will alter the density of the medium and hence the speed of propagating acoustic waves. Varying and indeterminate acoustic transduction adds another layer of difficulty to our efforts to model the acoustic environment for combating noise and reverberation.

The current solution is a communication cap-based audio (CCA) system. As shown in Fig. 12.1, astronauts wear such a cap under their helmets. It has redundant, built-in microphones and earphones. The use of close-talking, noise cancelling microphones of a CCA can dramatically help improve speech intelligibility, but only when the microphones are very close to the crewmember's mouth. Moving them away from the mouth, even though only for a very short distance, can cause great performance degradation. Such sensitivity gives arise to a number of recognized logistical issues and inconveniences that cannot be resolved with incremental improvements to the basic design of the CCA systems:

- the communication cap and the microphone booms cannot be adjusted during EVA operations (which last from four to eight hours) or during launch and entry,
- the microphones that are right next to the mouth can interfere in the suit subject from eating and drinking and, on the other hand, can easily be contaminated by food and drinks,
- since the communication cap needs to be individually tailored and the microphones need to be properly mounted according to the head size of each astronaut, caps in a number of different sizes have to be built and maintained (e.g., there are five sizes for Space Shuttle and International Space Station Extravehicular Mobility Unit CCA), and
- wire fatigue and blind mating of the connectors are also problems with the CCA.

Therefore a great effort is under way in NASA to develop a multiple-microphone audio system that is integrated with spacesuits and would be able to possess similar performance to a CCA. Innovations in beamforming and multichannel acoustic signal processing technologies are then solicited to improve in-helmet voice communication with better experiences. We (the research team of WeVoice, Inc.) were selected for a NASA Phase I SBIR (small business innovation research) award and were funded to carry out a 6-month pilot feasibility study in 2008.

12.1 Problem Description   185

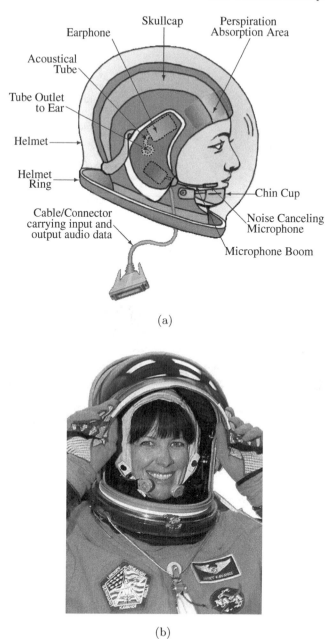

Fig. 12.1. Illustration of the current communication cap-based audio (CCA) system in a spacesuit helmet: (a) schematic of the CCA system and (b) an example of the CCA system worn by a female astronaut.

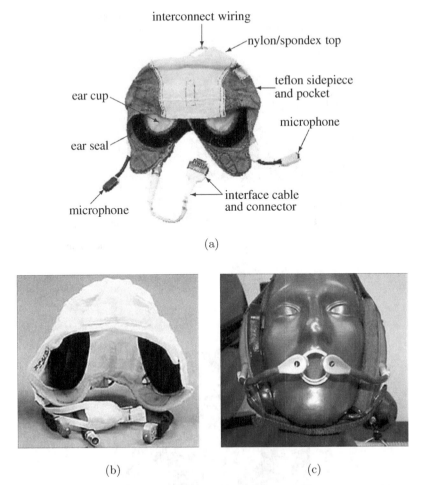

Fig. 12.2. Three currently used NASA communication caps: (a) shuttle/international space station extravehicular mobility unit (EMU) cap, (b) advanced crew escape suit (ACES) cap, and (c) developmental cap.

12.2 Problem Analysis

12.2.1 Sources of Noise in Spacesuits

There are two forms of noise in spacesuits: ambient noise that penetrates from the outside into the inside of the spacesuits, and suit structure-borne noise. The level of ambient noise is subject to a large range of variations, depending on the operation that the space shuttle/vehicle and the suit subject are undertaking. During launch, entry descent, and landing, the sound pressure level (SPL) of impulse ambient noise is < 140 dB. But during on-orbit and EVA operations, the surrounding environment of a crewmember is quiet. The

12.2 Problem Analysis   187

Table 12.1. Upper limits on continuous on-orbit noise level by frequency.

Band Center Frequency (Hz)	63 125 250 500 1k 2k 4k 8k 16k
Sound Pressure Level (dB)	72 65 60 56 53 51 50 49 48

Table 12.2. Sound pressure level (SPL) chart of typical noise fields.

SPL (dB)	Perception	Typical Environments
85 – 95	Very High Noise:	Construction Site
	Speech Almost	Loud Machine Shop
	Impossible to Hear	Noisy Manufacturing
75 – 85	High Noise:	Assembly Line
	Speech is Difficult	Crowded Bus/Transit Waiting Area
	to Hear	Very Noisy Restaurant/Bar
65 – 75	Medium Noise:	Department Store
	Must Raise Voice	Bank/Public Area
	to be Heard	Supermarket
55 – 65	Low Noise:	Doctor's Office
	Speech is Easy	Hospital
	to Hear	Hotel Lobby

upper limits on continuous on-orbit noise level by frequency are shown in Table 12.1. By comparing these SPL values to those of typical noise fields given in Table 12.2, we see that the crewmembers talk and listen as if they were at best in a doctor's office and at worst in a supermarket. Therefore, during EVA operations, ambient noise is at most a minor problem, and structure-borne noise is more imperative to be solved for in-helmet voice communication.

Four sources have been identified for the structure-borne noise in spacesuits [2]:

- airflow and air inlet hissing noise, as well as fan/pump noise due to required air supply and circulation,
- arm, leg, and hip bearing noise,
- suit-impact noise (e.g., footfall), and
- swishing-like noise due to air movement caused by walking (since the suits are closed pressure environments).

For head-mounted CCA systems, since the suit subject's body does not transmit bearing and impact noise, only airflow-related noise needs to be properly controlled. However, for an integrated audio system, structure-borne vibration

188   12 Practical Example

easily adds on the microphone outputs since the microphones are mounted directly on the suit structure.

12.2.2 Noise Cancelling Microphones

Noise cancelling microphones are nothing mysterious but close-talking differential microphone arrays. A differential microphone consists of an array of pressure sensors which are exposed to the incident sound. The microphone is therefore responsive to the spatial derivatives (gradients) of the scalar acoustic pressure field. Its output is a finite-difference approximation to the sum of those spatial derivatives. The response of a *first-order* differential microphone array (FODMA) is the combination of the zeroth-order signal and the first-order spatial derivative, while an ith-order ($i \geq 1$) array has a response proportional to a linear combination of signal derived from spatial derivatives up to, and including order i.

An FODMA, as depicted in Fig. 12.3(a), consists of two omnidirectional sensor elements with inter-element spacing d. The (zero-order) responses of the two omnidirectional microphones due to a sound source at a distance of r as a function of frequency f are

$$H_n(r_n; f) = \frac{e^{-j2\pi f r_n/c}}{r_n}, \quad n = 1, 2, \tag{12.1}$$

where

$$r_1 = \sqrt{r^2 - rd\cos\theta + d^2/4},$$
$$r_2 = \sqrt{r^2 + rd\cos\theta + d^2/4},$$

θ is the incident angle of the sound source with respect to the sensor axis, and c is the speed of sound whose value (in m/s) in air (similarly in pure oxygen, which is diatomic gas) can be calculated from the air temperature t_{air} (in degrees Celsius) using the following formula

$$c \approx 331.3 \times \sqrt{1 + \frac{t_{\text{air}}}{273.15}}. \tag{12.2}$$

Then the FODMA response can be written as

$$
\begin{aligned}
H_{\text{FODMA}}(r, \theta; f) &= H_1(r_1; f) - H_2(r_2; f) \\
&= \frac{e^{-j2\pi f r_1/c}}{r_1} - \frac{e^{-j2\pi f r_2/c}}{r_2}.
\end{aligned} \tag{12.3}
$$

Similarly the response of a second-order differential microphone array (SODMA), as shown in Fig. 12.3(b), is deduced as

$$H_{\text{SODMA}}(r, \theta; f) = \frac{e^{-j2\pi f r_1/c}}{r_1} - 2\frac{e^{-j2\pi f r_2/c}}{r_2} + \frac{e^{-j2\pi f r_3/c}}{r_3}, \tag{12.4}$$

12.2 Problem Analysis 189

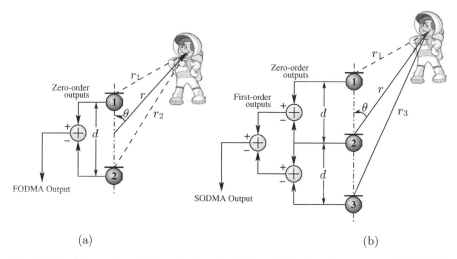

Fig. 12.3. Schematics of (a) a first-order differential microphone array (FODMA) and (b) a second-order differential microphone array (SODMA).

where

$$r_1 = \sqrt{r^2 - 2rd\cos\theta + d^2},$$
$$r_2 = r,$$
$$r_3 = \sqrt{r^2 + 2rd\cos\theta + d^2}.$$

Figure 12.4(a) plots the FODMA and SODMA responses to an on-axis sound source ($\theta = 0°$) evaluated at $r = 50$ mm and 1 m for $d = 10$ mm. The farfield responses indicate a highpass filter with a slope of $6i$ dB/octave for an ith-order ($i = 1, 2$) differential microphone array.

Consider two on-axis sound sources: one is the sound source of interest in the near field at a distance of r_s, and the other is a noise source in the far field at a distance of r_n. Then the gain in signal-to-noise ratio (GSNR) by using a differential microphone array (DMA) over using an omnidirectional microphone is found as

$$\text{GSNR}_{\text{DMA}} = 20\log_{10}\left\{\frac{|H_{\text{DMA}}(r_s, \theta_s; f)|}{|H_{\text{DMA}}(r_n, \theta_n; f)|}\right\} - 20\log_{10}\left(\frac{r_n}{r_s}\right), \quad (12.5)$$

where DMA can be either FODMA or SODMA. Figure 12.4(b) presents the GSNRs for $r_s = 15$ mm, $\theta_s = \theta_n = 0°$, and $d = 10$ mm. It is clearly demonstrated that differential microphones inherently suppress farfield noise. The higher the order of a DMA and the larger the distance of a noise source, the more attenuation is gained against the noise. Moreover, the farfield noise signal is more suppressed in the low frequencies than in the high frequencies.

While differential microphones are found very useful in situations where the background noise level is very high (like what exist in spacesuits), a well-

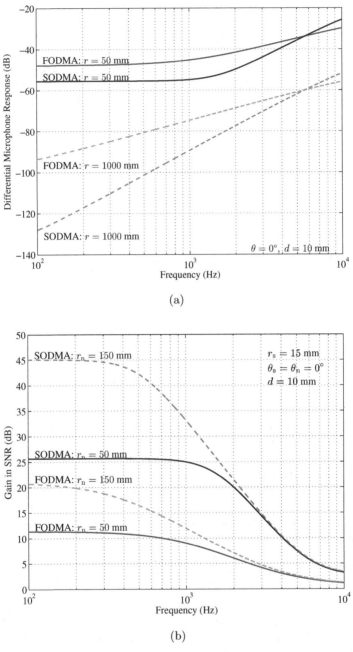

Fig. 12.4. Performance comparison of an FODMA and an SODMA for near-field speech acquisition. (a) Differential microphone array response and (b) gain in SNR over using an omnidirectional microphone.

12.2 Problem Analysis 191

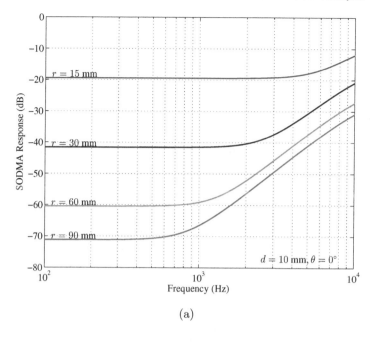

(a)

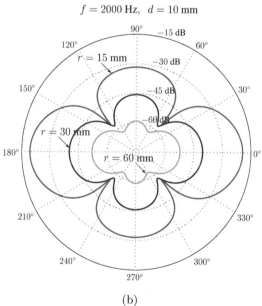

(b)

Fig. 12.5. Illustration of SODMA's sensitivity to (a) the range and (b) the incident angle of the near-field sound source of interest.

192   12 Practical Example

known, prominent drawback of differential microphone arrays is that their frequency response and level are extremely sensitive to the position and orientation of the arrays relative to the speaker's mouth. Figure 12.5 visualizes the sensitivity of an SODMA's response to r and θ. The inter-element spacing is $d = 10$ mm. Suppose that the microphone is initially placed 15 mm away from the mouth. When we move it only 45 mm (which is significantly less than an average distance between a man's mouth and one of his ears) further away, as seen from Fig. 12.5(a), the low-frequency speech components (between 100 Hz and 1 kHz) are attenuated more than 40 dB. At such a distance with such a level of attenuation, the captured speech signal becomes very weak and submerged in the background noise. Very poor speech intelligibility can then be expected and noise cancelling microphones no longer work as their name implies. In addition, we can see that the farther we move the microphone away from the mouth, the more attenuation is applied to the high-frequency components of the speech signal. This will not only reduce its intelligibility but also lower its quality. If the distance between the microphone and the mouth is fixed but the mouth deviates from the axis of the differential microphone, then from Fig. 12.5(b) we see that the response also changes remarkably. Since differential microphones are supposedly in close proximity to a user's mouth, a small displacement of the mouth can lead to a large change in the incident angle with respect to the array axis.

The sensitivity of a differential microphone to dislocation and disorientation makes it necessary to perform frequency and level equalization to its response according to the range and incident angle of its user's mouth [116]. However, this equalization will not affect the SNR and the directional response pattern. In the axis direction ($\theta = 0°$), the response is always the largest. In other words, the main lobe cannot be electrically steered to the mouth. Therefore, it can be concluded that noise cancelling microphones will not be used in the next-generation voice communication systems integrated with spacesuits.

12.3 Suggested Algorithms

We have learned that noise inside the helmet of a spacesuit has complicated characteristics. In addition, because of the presence of strong reverberation, the suit subject's speech and noise signals are mixed in a complex way in microphone outputs. To extract clean speech with high intelligibility and quality, noise reduction and speech enhancement will be explored in three domains: the time, frequency, and space domains. Hopefully a line can be drawn between noise and the interested speech signal. As illustrated by Fig. 12.6, we thought that there could be four techniques that are potentially helpful and deserve to be carefully studied: namely beamforming, multichannel noise reduction, adaptive noise cancellation, and single-channel noise reduction. As will be explained below, these algorithms (except beamforming and multichannel noise reduction) are not mutually exclusive, but in fact complementary and should

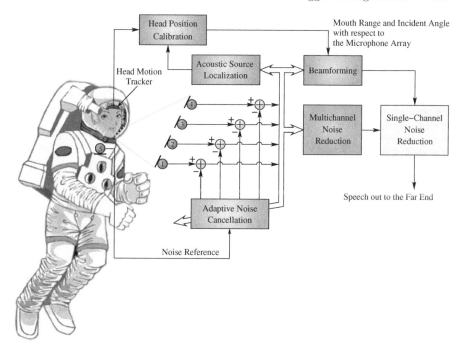

Fig. 12.6. Block diagram of the suggested algorithms to be used in the integrated spacesuit audio (ISA) system for in-helmet voice communication.

be used in concert. In general spatial cues are more vulnerable to improper processing than temporal/spectral cues. Consequently spatial processing will be carried out prior to time-frequency analysis.

12.3.1 Nearfield, Wideband Microphone Array Beamforming for Speech Acquisition in Spacesuits

Beamforming is a means of spatial filtering, which lets the signal coming from the pre-specified look direction pass through while suppressing signals propagating from other directions. The origin and the first implementation of the idea can be traced a long way back to a century ago. Nowadays one can find various beamformers in the literature. But their structures are not significantly different. The general structure of a beamformer is shown in Fig. 12.7. When the speech source of interest is in the far field of the microphone array, its corresponding time difference of arrival (TDOA) between the first and the nth ($n = 2, 3, \ldots, N$, where N is the number of microphones) microphone outputs τ_n is dependent only on the incident angle θ of the speech source. But when the speech source is in the near field, the TDOA τ_n depends on the position of the speech source $\mathbf{r}_s$.

194   12 Practical Example

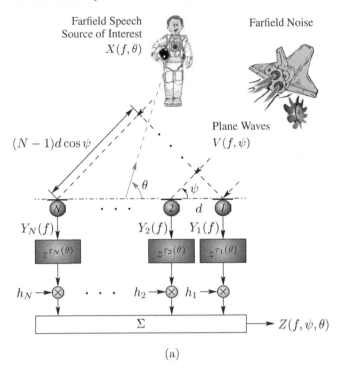

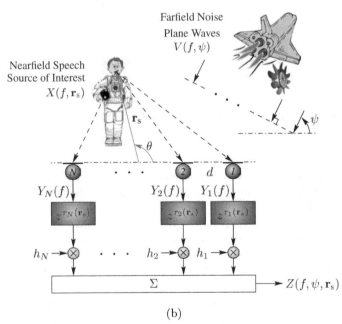

Fig. 12.7. General structure of a beamformer when the speech source of interest is in (a) the far field and (b) the near field of the microphone array.

There are two parameters to tune for a beamformer: the TDOAs $\tau_n(\mathbf{r}_s)$ and the filter coefficients h_n. The TDOAs specify the look direction (or more precisely the look position in the nearfield case) and the filter coefficients control the beam pattern of the array: the beamwidth of the mainlobe, the characteristics of the sidelobes, and the position of the nulls. For a delay-and-sum beamformer (known as the classical beamforming algorithm), $h_n = 1/N$ ($\forall n = 1, 2, \ldots, N$). As a result, speech signal originating from the desired direction are summed in phase, while other signals undergo destructive interference. If noise is incoherent, the improvement in SNR would be $10 \log_{10} N$ dB. When N is small, only a very limited gain in SNR can be obtained, which is insufficient and practically not so useful. For coherent noise, the performance of the delay-and-sum beamformer is strongly dependent on the direction of arrival of the noise signal. If the direction of noise coincides with that of speech signal, no SNR gain can be possibly produced. Since the microphone weights of the classical delay-and-sum beamformer are constants and frequency independent, its spatial response is a function of frequency, as shown in Figs. 12.8 and 12.9. As the frequency increases, the beam becomes commensurately narrower. This implies that when the speech source deviates from the look direction, speech distortion will be observed in the beamformer output. The high-frequency components are more attenuated than the low-frequency components, i.e., the speech signal will be low-pass filtered.

In order to have a constant-beamwidth spatial response over a wide frequency range of speech spectrum, harmonically-nested subarrays [51], [52], [83] and the more generalized filter-and-sum beamformer [77], [115] were proposed. Harmonically-nested subarrays are large in size and require a great number of microphones. For integrated spacesuit audio (ISA) systems, a filter-and-sum beamformer is practically more useful. In the design of a *fixed* filter-and-sum beamformer, the microphone weights can be computed using a least-squares (LS) filter design method [94], [95], which can yield satisfactory results [15] as exemplified by the far-field response of a filter-and-sum beamformer visualized in Fig. 12.10 in comparison with that of a delay-and-sum beamformer previously shown in Fig. 12.8. But when a nearfield speech source is concerned, the LS algorithm was found not as stable as an LCMV (linearly constrained minimum variance) algorithm.

Spatial responses of fixed beamformers are static since the microphone weights, once designed, are fixed. Fixed beamformers are data-independent spatial filters. During the design of a fixed beamformer, noise field is not known and the isotropic model that is a reasonable first-order approximation of most of the real noise fields is commonly used. If the assumption matches the acoustic condition in which the array is used, this approach leads to a good beamformer design. However, such a simple model can never be real in practice and therefore a fixed beamformer is at best suboptimal even though it can be steered to accurately follow the movement of the speech source. For example, if noise is coming from a point sound source, ideally the beamformer

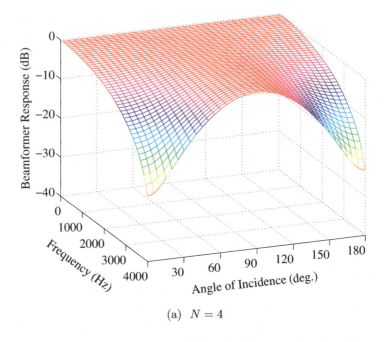

(a) $N = 4$

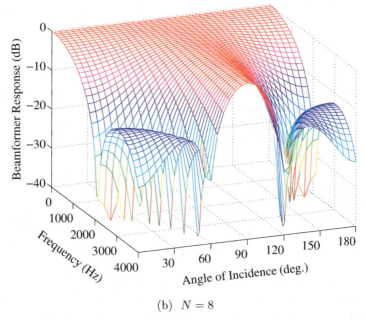

(b) $N = 8$

Fig. 12.8. Response of a delay-and-sum beamformer for two equally spaced linear arrays containing respectively (a) $N = 4$ and (b) $N = 8$ microphones spaced by $d = 20$ mm and with the look direction $\theta = 90°$. The sound sources are assumed in the far field.

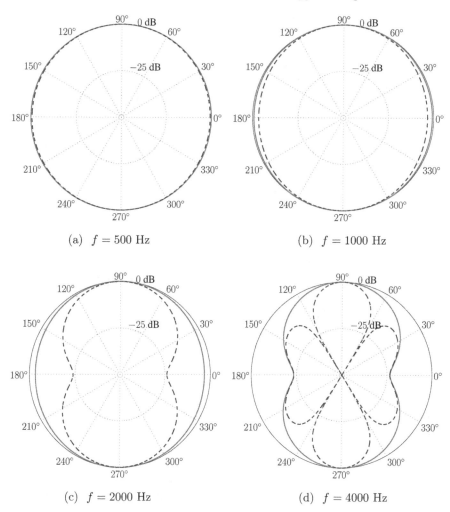

Fig. 12.9. Directional patterns of a delay-and-sum beamformer for two equally spaced linear arrays containing respectively $N = 4$ (red solid line) and $N = 8$ (blue dashed line) microphones spaced by $d = 20$ mm and with the look direction $\theta = 90°$. The sound sources are assumed in the far field.

should be able to put a null in that direction. However, only by chance a fixed beamformer can do so.

In this line of thoughts, adaptive beamformers were proposed. Adaptive beamformers try to track the variation of the surrounding noise field and adaptively search for the optimum location of the nulls that can most significantly reduce noise under the constraints that the interested speech signal is not distorted at the beamformer output. This is then formulated as the widely-known linearly constrained minimum variance (LCMV) problem. The

198   12 Practical Example

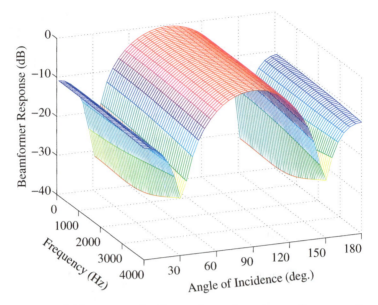

Fig. 12.10. Far-field response of a filter-and-sum beamformer for an equally spaced linear array containing $N = 4$ microphones spaced by $d = 20$ mm and with the look direction $\theta = 90°$. In the design of such a beamformer with the least-squares method, the width of the mainlobe was set as $40°$.

difficulty of implementing an adaptive beamformer is how to properly define the constraint. In an anechoic environment, only the direction of the speech source needs to be known and the LCMV solution leads to the minimum variance distortionless response (MVDR) beamformer due to Capon [21]. But in a real reverberant environment, the linear constraint of the LCMV algorithm requires the knowledge of the impulse responses from the speech source to each microphone [13]. The LCMV beamformer was first developed by Frost [53] and is also called the Frost beamformer. One variant of the Frost beamformer is the so-called generalized sidelobe canceller (GSC) proposed by Griffiths and Jim [61]. The GSC transforms the LCMV algorithm from a constrained optimization problem into an unconstrained form. Therefore, the GSC and LCMV beamformers are essentially the same while the GSC can lower the computational cost by forcing the constraint into the front-end of the array processing [79], [123].

The LCMV beamformer is theoretically appealing since it seems to be an optimal spatial filter. But in practice, when an MVDR beamformer is used in a reverberant environment or when the estimates of the acoustic impulse responses in the Frost/GSC beamformer have errors (which is certainly inevitable), the so-called signal cancellation problem would occur and only a suboptimal performance can be achieved.

12.3 Suggested Algorithms   199

Our perception of speech remarkably benefits from spatial hearing since it can be easily experienced by observing the difference in understanding between using both ears and with either ear covered when listing in an enclosed space where there are multiple (both speech and noise) sound sources at the same time. This is a striking capability since we accomplish it with only two ears and without much efforts. Consequently some people may mistakenly believe that a small array of only two microphones would be able to give us quite some gain in SNR via beamforming. One needs to understand that spatial filtering is only a part, rather than the entirety, of the cocktail-party effect of human auditory systems. With a man-made microphone array system, we summarize that

- for incoherent noise sources, the SNR gain is low if a small number of microphones are used;
- for coherent noise sources whose directions are different from that of the speech source, a theoretically optimal gain in SNR can be high but is difficult to obtain due to practical limitations (e.g., unavailability of a precise a priori knowledge of the acoustic impulse responses from the speech sources to the microphones, inconsistent responses of the microphones across the array, etc.); and
- for coherent noise sources that are in the same direction as the speech source, beamforming (as a spatial filter) is ineffective.

One way to achieve a higher noise reduction gain at a price of some speech distortion is to replace the *hard* constraint in the LCMV beamformer with a *soft* constraint [81], [113], which becomes nonlinear and allows for some deviation in magnitude and phase for signals propagating from the look direction. The other way is to have a post-filter at the beamformer output [80], [127], [128]. The idea of sequential beamforming and post-filtering is based on the discovery that the multichannel Wiener filter, which minimizes the MSE in the context of speech extraction with a microphone array, can be factorized as a product of the LCMV beamformer and a single-channel Wiener filter for noise reduction [111]. The former approach involves a nonlinear optimization problem and a gradient search has to be employed. In addition, the amount of speech distortion is hard to adjust. It is the post-filtering technique which has attracted an increased interest and has achieved some successes. In Section 12.3.3 that follows, we will discuss various single-channel noise reduction algorithms for post-filtering.

An essential requirement of microphone array beamformers, either fixed or adaptive, is the knowledge about the position of the suit subject's mouth. This can be met by using acoustic source localization and tracking method or infrared marker-based 3D motion tracking systems. While acoustic methods can make the whole system more compact with no need to add additional hardware, their level of precision is not in the same order of magnitude as that of marker-based 3D motion tracking systems, not even to mention the reliability in the face of strong noise and reverberation. A small marker will be

200   12 Practical Example

placed on the suit subject's head and would presumably cause little distraction or logistical inconvenience. The marker-based motion tracking system provides the marker's position relative to the system's origin. This origin needs to be aligned to the origin of the microphone array. Otherwise the data of the mouth position cannot be used. The alignment can be manually carried out, but an automatic way with the help of acoustic source localization techniques is by all means more preferable. The calibration is performed only once when the spacesuit is put on or when the audio system is rebooted. During this operation mode, the acoustic condition can be well controlled[1] and a large amount of acoustic data can be collected. Therefore a high degree of accuracy of acoustic source localization algorithms can be expected. For the state of the art of time delay estimation and acoustic source localization, the interested reader can refer to a recent survey [73] and the references therein.

12.3.2 Multichannel Noise Reduction: a More Practical Microphone Array Signal Processing Technique

Performance degradation of an adaptive beamformer in reverberant environments is a well-known problem due to the unavailability of an accurate *a priori* knowledge of the acoustic channel impulse responses from the speech source to each microphone. As clearly explained in the previous chapters, a beamformer has two effects: dereverberation and denoising. The performance degradation of an adaptive beamformer in reverberant environments is mainly caused by poor dereverberation results [10], [15]. If we cannot do a good job with a beamformer in dereverberation, the better strategy may be not to do it at all. Therefore the reverberant speech signal at one arbitrary microphone across the array, instead of the original speech source signal, is what to be estimated from the multiple microphone outputs [26], [74]. This suggests that multichannel noise reduction should be used.

In the so-called transfer function generalized sidelobe canceller (TF-GSC) algorithm proposed by Gannot et al. [55], [56], transfer function ratios with respect to various microphone pairs (instead of transfer functions themselves in traditional adaptive beamforming algorithms) are *blindly* estimated by exploiting the non-stationarity of speech. But this only partially takes advantage of the idea of multichannel noise reduction. In [15], we showed a trick which theoretically eliminates the need to estimate the transfer function ratios, which leads to a more practical approach to reducing noise with multiple microphones.

Another convenience of using a multichannel noise reduction algorithm compared to a traditional adaptive beamformer is that the array geometry and the sensor consistency in gain and frequency response are not critical. This makes the design and fabrication of the microphone array in spacesuits much easier and more flexible.

---

[1] For example, air supply hasn't been turned on yet, the suit subject stands still, and even the helmet hasn't been put on the suit.

In the traditional microphone array processing, the original speech signal is to be recovered and the Wiener filter (which is the optimal solution in the MMSE sense) can be factorized into the product of an LCMV beamformer and a post-filter [111]. For multichannel noise reduction, the reverberant speech signal received at one microphone across the array is the goal and the Wiener-filter solution again can be factorized into the product of an LCMV filter and a post-filter [15]. While the LCMV filters are different in the two approaches, the post-filters are identical. Therefore the single-channel noise reduction methods to be discussed below are shared by both approaches.

12.3.3 Single-Channel Noise Reduction

While microphone array beamforming or the multichannel noise reduction technique can effectively attenuate noise theoretically without introducing much distortion to speech, the achievable gain in SNR is limited by the number of microphones to be employed. If that number is small due to the size of the helmet, the residual noise can still be quite strong. Both speech distortion and additive noise impair speech intelligibility. So it is not an optimal solution, in terms of speech intelligibility, to keep speech distortion very low while allowing strong residual noise. Instead, attaining a balance between speech distortion and noise level is what we desire for. Single-channel noise reduction methods boost the SNR but inevitably incur speech distortion. They essentially trade off speech distortion for noise reduction, and therefore can be ideally used as a post-filtering step for microphone arrays.

So far a large number of single-channel noise reduction algorithms have been invented and they can be broadly classified into three classes: the class of filtering techniques, the class of spectral restoration methods, and the class of speech model-based approaches. The first two classes of methods have been extensively studied in this book while typical algorithms in the third class include harmonic-model-based methods [85], linear-prediction-model-based Kalman filtering approaches [97], and hidden-Markov-model-based statistical methods [46].

While the filtering and spectral restoration methods tackle the noise reduction problem from different perspectives, their implementations are all based on the estimates of second-order statistics (SOS) of the noisy speech and noise signals. Accurate and timely estimation of these statistics is crucial to the success of these algorithms. A common way to estimate the SOS of noise is to implement a voice activity detector (VAD) and then perform the estimation when speech is absent [28], [40], [89], [104]. In order to ensure that the noise SOS estimated during silent periods are useful when speech is present, a stationary or slowly varying noise signal has to be assumed. If noise unfortunately varies fast in time, none of the filtering and spectral restoration algorithms can work well. Another way is to directly estimate the SOS of the clean speech signal from multiple microphone outputs. Apparently this method can only be used when the developed noise reduction algorithm serves as a post-filter

202   12 Practical Example

of a microphone array system. With this speech spectral estimation method, noise has to be *incoherent* at the multiple microphones [127], which limits its use in practice.

12.3.4 Adaptive Noise Cancellation

As clearly explained above, a nonstationary noise signal is in general difficult to reduce due to the lack of the ability to instantaneously track its time-varying statistics. Good examples inside spacesuits are spacesuit bearing noise and footfall. They are short bursting and randomly occur. If they have no overlap in time with the suit subject's speech, we can develop a sophisticated detector to detect noise transients and actively eliminated them. The detection criterion can be based on a single-channel microphone output or multichannel microphone outputs [29]. However, bearing/footfall noise and speech can be simultaneously observed. In these cases, only adaptive noise cancellation (ANC) can possibly produce some good results.

Among the four noise sources summarized previously in Section 12.2.1, spacesuit bearing noise and suit-impact noise signals are transmitted by the spacesuit from the lower parts of the suit to the microphones around the helmet. If we have a microphone woven in the inner fabric of the suit at a place in the upper torso where the suit subject's speech is inaudible, the sensor will only pick up vibrations that propagate through and its output can be nicely used as a noise reference signal. Then we can use ANC and remove the components that are linearly correlated to the noise reference signal from the multichannel microphone array outputs. If the beamformer is fixed, it makes no difference in performance (without taking into account the complexity) whether ANC or beamforming is first performed. But if an adaptive beamformer is employed, the order is definitely critical and the favor goes to ANC being first performed, as depicted in Fig. 12.6.

In ANC, a reference noise signal is assumed available, as illustrated by Fig. 12.11. This reference signal $v_2(k)$ is used to estimate the noise $v_1(k)$ in the primary microphone output $y(k)$ by a linear filter, and this estimate $\hat{v}_1(k)$ may then be subtracted from $y(k)$ to form an estimate of the speech signal $x(k)$. If $x(k)$, $v_1(k)$, and $v_2(k)$ are jointly wide-sense stationary processes, then a Wiener filter may be designed to find the MMSE estimate of $v_1(n)$. In practice, however, a stationarity assumption is not generally appropriate. Therefore, as an alternative to the Wiener filter, an adaptive filter is used.

Needless to emphasize, the key to the successful operation of the adaptive noise canceller is the availability of a good reference signal $v_2(k)$ which has a strong correlation with $v_1(k)$ but ideally has no correlation with the speech signal $x(k)$. The performance of the employed adaptive algorithm is also critical. If the adaptive algorithm cannot converge fast enough to track time variation of the noise signals, the estimate of the noise signal cannot be reliable. The problem that the adaptive filter faces in ANC is similar to that in acoustic echo cancellation (AEC). Thereafter a number of recently

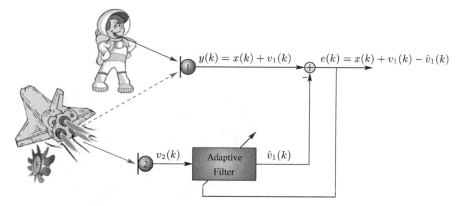

Fig. 12.11. Illustration of the concept of adaptive noise cancellation when a reference noise signal $v_2(k)$ is available.

proposed fast adaptive algorithms that were found useful for AEC can also be helpful for ANC. They include the frequency-domain adaptive filters [4], the improved proportionate NLMS (IPNLMS) algorithm [5], and the exponentiated gradient (EG) algorithm [7].

12.4 Algorithm Validation

12.4.1 In-Helmet Multichannel Acoustic Data Collection

For algorithm analysis and validation, two sets of multichannel acoustic data were collected from inside of several spacesuits at the beginning of this research. One set of the data was measured by us (i.e., WeVoice, Inc.) and NASA engineers from the Glenn Research Center (GRC) and Johnson Space Center (JSC) in March 2008. The other was measured by Begault and Hieronymus from the NASA Ames Research Center (ARC) in early 2007 [2].

WeVoice-GRC Experimental Setup

In the WeVoice-GRC data collection effort, two NASA's current prototype planetary exploration suits were used. They are the Mark III (a joint effort between Air-Lock, Inc. and ILC Dover, Inc.) and the REI (Rear Entry I-Suit developed and fabricated by ILC Dover, Inc.) suits, as shown in Fig. 12.12. Both are rear-entry suits, unlike the EMU currently in use, which is a waist-entry suit. The Mark III incorporates a hybrid mixture of hard and soft suit components, including hard upper torso (HUT), hard brief and hip/thigh elements made of graphite/epoxy composite, bearings at the shoulder, upper arm, hip, waist, and ankle, and soft fabric joints at the elbow, knee, and ankle, and a pair of military flight boots. The I-Suit is primarily a soft suit, yet

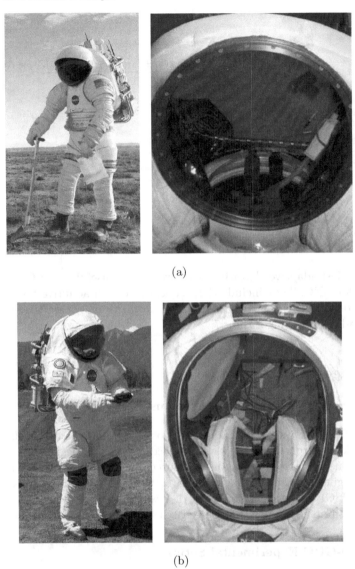

Fig. 12.12. Comparison of two NASA's prototype planetary exploration spacesuits: (a) the Mark III with a round helmet neck ring and (b) the REI (Rear Entry I-Suit) with an oval ring.

incorporates a limited number of bearings at the wrist, shoulder, upper arm, upper hip, upper leg joints (the other joints on this suit are fabric joints). In addition, it has a pair of mountaineering boots. The I-Suit represents a compromise between an HUT suit and an all soft suit, such as the Apollo A7LB suit. As illustrated in Fig. 12.12, the Mark III has a round helmet neck

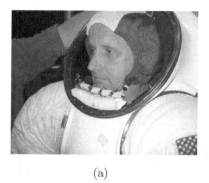
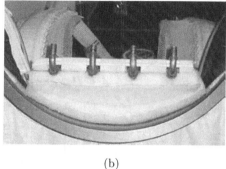

(a)   (b)

Fig. 12.13. Illustration of two equispaced, linear microphone arrays and their mounting apparatuses developed in the multichannel acoustic data collection experiments. (a) The SMS array mounted in a Mark III suit and (b) the Pro-Audio microphone array mounted in a REI suit.

ring while the REI has an oval ring. The Mark III is much (about 50 pounds) heavier but has better mobility than the I-Suit [117].

Two types of microphone array and acoustic data acquisition systems were used. One is an SPL (Sound Pressure Level) Measurement System (SMS) and the other is a digital Pro-Audio system. Each system has its own 4-element linear microphone array with 4 cm inter-element spacing located in front of the suit subject. Specialized microphone array mounting apparatuses were developed at NASA-JSC, as illustrated in Fig. 12.13. In addition to the 4-element array, a fifth microphone (of the same type as that used in the array) was positioned on the rear panel (for the study on adaptive noise cancellation) and a close-talking microphone was also used. The rear panel on the Mark III suit had an acoustic foam panel covering the rear where the microphone was located.

The SMS provided by the NASA GSC EVA Audio Team is a NIST-calibrated acoustic data collection system. Its array consists of 4 G.R.A.S. Type 40BE pre-polarized, free-field instrumentation microphones with type 26CB preamp modules. An NI-4722 24-bit DAQ performs analog-to-digital conversion (ADC) and provides the IEPE current supply for the microphone preamps. The Pro-Audio system was designed by WeVoice, Inc. It employs an array of 4 MWM Acoustics EM6022 omni-directional electret microphone, a Grace M802 microphone preamp with digital audio outputs, a Rosetta 800 Apogee audio interface, and a PC Laptop (Windows XP) with Adobe Audition software. The block diagrams of these two multichannel acoustic data acquisition systems are presented in Fig. 12.14.

In the experiments, the Mark III and REI suits were fitted with both the SMS and the Pro-Audio microphone arrays. An extensive set of tests were run under three measurement conditions: 1) subject standing, 2) subject walking slow, and 3) subject walking fast. The experiments were conducted

206   12 Practical Example

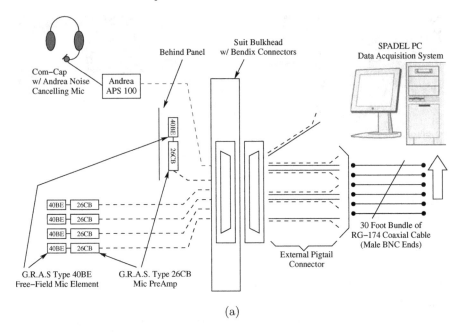

(a)

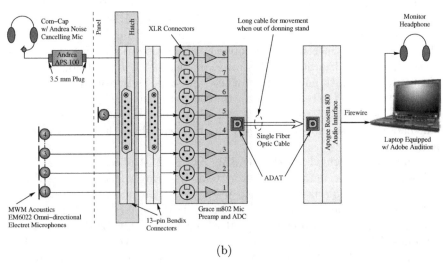

(b)

Fig. 12.14. Block diagrams of (a) the SPL measurement and (b) the digital Pro-Audio multichannel acoustic data acquisition systems.

using either a Liquid Air Backpack (LAB) or a K-Bottle based air supply. The K-Bottle supply simulates an umbilical configuration while the LAB simulates a PLSS (Primary, or Portable, Life Support System/Subsystem) based configuration. Over 11 gigabytes of data were collected.

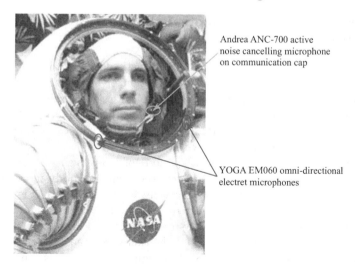

Fig. 12.15. Mark III suit with helmet bubble removed showing a head-mounted Andrea ANC-700 active noise cancelling microphone on the communication cap and two Yoga EM060 electret microphones as placed along the helmet ring. (Courtesy Durand R. Begault and James L. Hieronymus, NASA ARC.)

ARC Experimental Setup

In the experiments that Begault and Hieronymus conducted [2], data were collected by a digital Pro-Audio system in a Mark III suit with LAB. While different microphone array, preamp, and ADC were used, their system architecture is similar to that of the WeVoice Pro-Audio system as illustrated in Fig. 12.14(b). Attached to the communication cap was also an Andrea ANC-700 active noise cancelling microphone used as a baseline reference. But the array consisted of only two YOGA EM 060 omni-directional electret microphones, which were positioned around the helmet-mounting ring of the suit on foam rubber mounts at 7 and 4 o'clock, respectively (the top of the ring is 12 o'clock and the bottom 6 o'clock), as shown in Fig. 12.15. The subject read a set of spoken dialogue system commands with the suit pressurized in the following 5 conditions:

1) subject inside suit on the "donning stand," a device that holds the suit in place,
2) subject standing,
3) subject walking (footfall impacts and hip bearing noise),
4) subject walking and moving arms (shoulder bearing noise plus walking noise), and
5) subject seated in a rover seat (volume change and head lower in helmet area).

208   12 Practical Example

In-Helmet Multichannel Acoustic Database and Data Preprocessing

Among the huge amount of multichannel acoustic data, not all the recordings are immaculate and can be used for validation of beamforming and multichannel noise reduction algorithms. Only typical, relevant recordings of good quality are selected to form a database for future use. A description of the database is presented in Table 12.3. All these multichannel sound clips were sampled at 48 kHz and quantized with 16 bits.

Shown in Fig. 12.16(a) is a period of waveform extracted from *grc_stand.wav* for which the suit subject stood with no body movement and hence only air flow related noise was present. Clearly seen from Fig. 12.16(c), the recording contained a loud low-frequency sound, which is similar to rumble noise (rumble noise is a loud low-frequency sound and is usually caused by ball bearings). A high-pass filter (HPF) therefore needs to be used to preprocess the microphone outputs. Such a high-pass filtering is critical to the beamforming and multichannel noise reduction algorithms that follow. Multichannel algorithms that exploit spatial sampling generally can yield just a theoretically marginal gain in speech enhancement for low-frequency signals. If a loud low-frequency noise signal dominates in the microphone output like what is shown in Fig. 12.16(a), then the full-band improvement of those multichannel algorithms will not manifest even though they can do a pretty good job in high-frequency bands. The processed signal using an HPF with the cutoff frequency of 100 Hz is shown in Fig. 12.16(b) and a comparison of spectrum of the original and processed noise signals is presented in Fig. 12.16(c).

Summarized in Table 12.4 are the SNRs of the recorded multichannel signals and the signals after being HP filtered. These SNRs will be used as the baseline references to assess the beamforming and multichannel noise reduction algorithms. Since speech is non-stationary in nature, estimating the SNR of a noise-corrupted speech signal is *not* a theoretically trivial problem. Sophisticated speech-model-based methods (e.g., hidden Markov model-based approaches) are computationally intensive, while simple algorithms may not be always reliable. In our research, a frame-based, histogram analysis algorithm is used for SNR estimation of speech signals. This algorithm divides a sufficiently long sequence of speech samples into frames and computes the energy of each frame. The average of the top 20% of the frame energies is regarded as the speech energy and the average of the bottom 20% as the noise energy. Their ratio leads to the SNR estimate of the signal.

12.4.2 Performance Evaluation of Beamforming Algorithms

In this study, fixed beamformers targeted for a nearfield speech sound source were first investigated. But unfortunately these fixed beamformers produced only marginal, if not nonexistent, gains in SNR. For these results, two possibilities can be surmised: 1) the speech source and the dominant noise sources

Table 12.3. Description of the multichannel acoustic database collected from inside spacesuits.

DAQ System	Spacesuit	Air Supply	File Name	Measurement Condition	Duration (s)
WeVoice (Pro-Audio)	Mark III	LAB	wv_stand.wav	standing	6
			wv_walkslow	walking slow	5
			wv_walkfast	walking fast	2
GRC (SMS)	Mark III	LAB	grc_stand.wav	standing	60
			grc_walkfast	walking fast	82
ARC (Pro-Audio)	Mark III	LAB	arc_donning_lookfwd.wav	donning stand & looking forward	23
			arc_donning_lookright.wav	donning stand & looking right	22
			arc_donning_lookleft.wav	donning stand & looking left	21
			arc_stand_lookfwd.wav	standing & looking forward	22.5
			arc_stand_lookright.wav	standing & looking right	24
			arc_stand_lookleft.wav	standing & looking left	23.5
			arc_walk.wav	walking	30
			arc_walk_movearm1.wav	walking & moving arms	26.5
			arc_walk_movearm2.wav		25.8

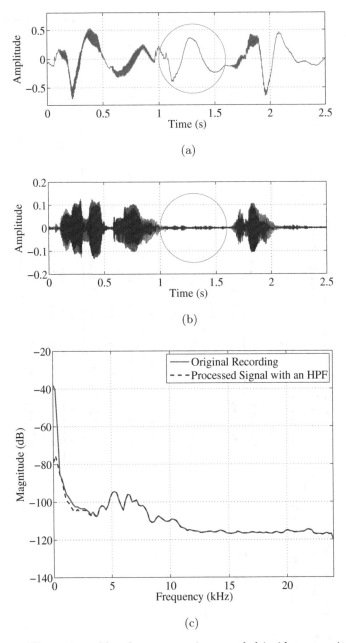

Fig. 12.16. Illustration of low-frequency noise recorded inside spacesuits and removing it with a high-pass filter (HPF). (a) A short period of waveform extracted from *grc_stand.wav*, (b) the processed signal using a high-pass filter with the cut-off frequency of 100 Hz, and (c) the comparison of spectrum between the original and the processed noise signals (corresponding to the circled parts).

Table 12.4. SNRs of the recorded multichannel signals and the signals after being high-pass filtered.

	SNR (dB)							
	Microphone Outputs				HP-Filtered Signals			
File Name	#1	#2	#3	#4	#1	#2	#3	#4
wv_stand.wav	21.309	19.683	19.567	21.508	22.891	22.393	22.012	22.335
wv_walkslow.wav	19.062	18.606	17.942	19.717	20.496	20.784	20.997	20.553
wv_walkfast.wav	19.431	19.028	19.061	19.251	20.265	20.750	20.686	19.755
grc_stand.wav	15.417	15.417	16.145	15.614	21.912	21.912	20.385	21.143
grc_walkfast.wav	17.067	15.541	15.590	16.216	17.766	18.027	17.207	18.094

	SNR (dB)			
	Microphone Outputs		HP-Filtered Signals	
File Name	#1	#2	#1	#2
arc_donning_lookfwd.wav	16.874	17.007	28.902	27.959
arc_donning_lookright.wav	18.560	18.152	29.507	27.300
arc_donning_lookleft.wav	18.066	18.328	28.618	28.744
arc_stand_lookfwd.wav	14.950	14.558	26.601	26.012
arc_stand_lookright.wav	17.593	16.829	26.881	25.933
arc_stand_lookleft.wav	16.853	16.292	27.646	27.383
arc_walk.wav	17.247	17.800	20.888	20.869
arc_walk_movearm1.wav	13.539	12.743	20.427	20.111
arc_walk_movearm2.wav	12.380	12.953	19.146	18.554

are in the same direction with respect to the array, 2) the assumption that the suit subject's mouth can be treated as a point sound source is problematic inside the helmet.

Recall that in the WeVoice and GRC measurements, the two arrays were mounted under the suit subject's jaw (see Fig. 12.13). Consequently, both the subject's mouth and the airflow inlet are normal to the array axis and the first surmise above is very likely true. Presented in Fig. 12.17 is a comparison of the magnitude of coherence between the signals in two periods of recording extracted from *wv_stand.wav*: one when speech was present and the other when there was only noise. It is evident that the noise signals at the four microphones are fairly coherent.

We studied the second surmise via a number of time delay estimation (TDE) algorithms for acoustic source localization. They included the full-band GCC-PHAT (generalized cross correlation-phase transform) method [84], the

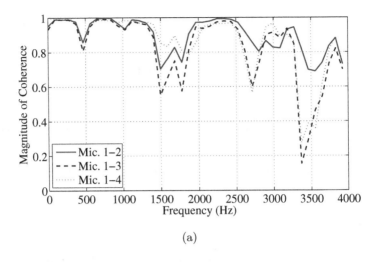

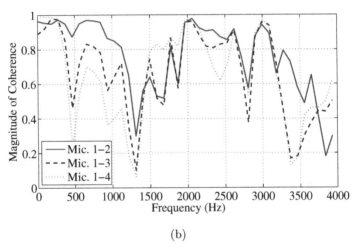

Fig. 12.17. Comparison of the magnitude of coherence between the signals in two periods of recording extracted from *wv_stand.wav*: (a) when speech was present and (b) when there was only noise.

subband GCC-PHAT algorithm, the adaptive eigenvalue decomposition decomposition algorithm [3], [70], the blind SIMO system identification based approach [71], and the spatial multichannel correlation technique [9], [23]. A voice activity detector was used and flawed detections around ambiguous speech-noise boundaries were manually corrected. But still none of the TDE algorithms can produce satisfyingly consistent and accurate results. While these can be mainly attributed to strong reverberation and loud noise inside

12.4 Algorithm Validation 213

Table 12.5. Summary of the simulation results of an adaptive subband MVDR beamformer (MVDR-BF) followed by a Wiener filter for single-channel noise reduction (WF-SCNR) with the multichannel acoustic data gathered in spacesuits.

File Name	Input Microphone SNR (dB)				Output SNR (dB) MVDR-BF	WF-SCNR	Average Gain (dB)
	#1	#2	#3	#4			
wv_stand.wav	22.89	22.39	22.01	22.33	22.97	34.62	12.21
wv_walkslow.wav	20.50	20.78	21.00	20.55	21.25	31.29	10.58
wv_walkfast.wav	20.27	20.75	20.69	19.76	21.47	31.45	11.08
grc_stand.wav	21.91	21.91	20.38	21.14	22.04	32.11	10.77
grc_walkfast.wav	17.76	18.02	17.21	18.09	17.97	26.64	8.87

File Name	Input Mic. SNR (dB)		Output SNR (dB) MVDR-BF	WF-SCNR	Average Gain (dB)
	#1	#2			
arc_donning_lookfwd.wav	28.91	27.96	29.55	42.74	14.31
arc_donning_lookright.wav	29.49	27.30	28.50	41.80	13.41
arc_donning_lookleft.wav	29.02	28.74	29.44	42.02	13.14
arc_stand_lookfwd.wav	26.59	26.02	26.92	39.38	13.07
arc_stand_lookright.wav	26.90	25.93	26.37	38.95	12.54
arc_stand_lookleft.wav	27.65	27.39	27.49	40.60	13.08
arc_walk.wav	20.90	20.87	21.97	32.08	11.20
arc_walkarm1.wav	20.42	20.11	21.22	32.34	12.07
arc_walkarm2.wav	19.14	18.56	18.65	29.43	10.58

the helmet, a negative impact caused by the problematic assumption that the mouth can be treated as a point sound source (by observing its size with respect to the distance from the array) is certainly unable to be eliminated.

In addition to fixed beamformers, two adaptive beamforming algorithms were also tested: the full-band and subband MVDR beamformers. The speech source was believed to be in the broadside. The subband MVDR beamformer performed better than the full-band implementation. The subband MVDR results are presented in Table 12.5. This study suggests that we need to move the microphone array to the side of the suit subject (such that the mouth and the airflow inlet could be in different directions to the array) in the following-up efforts for the development of ISA systems.

214   12 Practical Example

12.4.3 Validation of Multichannel Noise Reduction Algorithms

The simulation results of the subband MVDR algorithm for multichannel noise reduction (MCNR) are summarized in Table 12.6. By comparing these results to that presented in Table 12.5, it is very clear that the MVDR for MCNR performs better than the MVDR beamformer. In each set of the three measurements, the performance of both the beamformer and MCNR degrades as the subject increases movement and the noise becomes more non-stationary. Figure 12.18 visualizes the non-stationarity of the noise signals recorded under different conditions. When the suit was on the donning stand, the noise signal was at a low level and reasonably stationary. But as the subject's movement increased, the captured noise became more and more non-stationary. When the subject walked and moved his arms, the noise sounds like sanding wood. In this case, noise reduction would be very challenging.

12.4.4 Validation of Single-Channel Noise Reduction Algorithms

Tables 12.5 and 12.6 also include the performance of a Wiener-filter single-channel noise reduction algorithm as a post filter following the subband MVDR beamformer and the subband MVDR multichannel noise reduction, respectively. One can see that the Wiener filter for single-channel noise reduction we developed in this research is relatively robust and can reliably produce about 12 dB gain in SNR in all studied measurement conditions. In our experiments, the algorithm was tuned to yield such a level of gain with the amount of speech distortion that caused little intelligibility loss according to our informal listening tests. If more speech distortion is allowed, more gain in SNR can be achieved. Figure 12.19 visualizes the noise reduction procedure and plots the waveforms of the first microphone, the MCNR, and the SCNR outputs for the recording wv_stand.wav. The background airflow noise is clearly minimized.

12.4.5 Feasibility Assessment of Using Adaptive Noise Cancellation in Spacesuits

As previously explained, the key to the success of applying an adaptive noise cancellation algorithm inside spacesuits is that the reference noise signal is highly correlated with the noise components, but is little correlated with the speech components of the primary microphone outputs. These two conditions are equally important. By examining the reference microphone (Microphone 5 in the WeVoice and GRC measurements) signals, we found that they contain very evident speech signals from the suit subject, as seen from a sample of Microphone 5 outputs in comparison with the first microphone output shown in Fig. 12.20. Therefore, adaptive noise cancellation cannot be employed with the current installation of Microphone 5. In the future, Microphone 5 will be moved further away from the spacesuit helmet, ideally closer to the spacesuit bearings.

12.4 Algorithm Validation 215

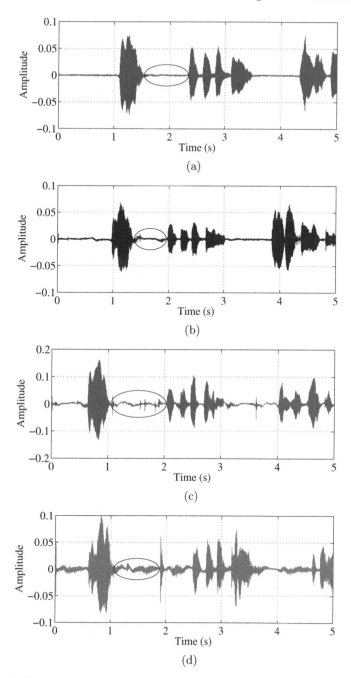

Fig. 12.18. Comparison of stationarity among the noise signals in the ARC measurements under different conditions: (a) subject inside suit on the "donning stand" (looking forward), (b) subject standing (looking forward), (c) subject walking, and (d) subject walking and moving arms.

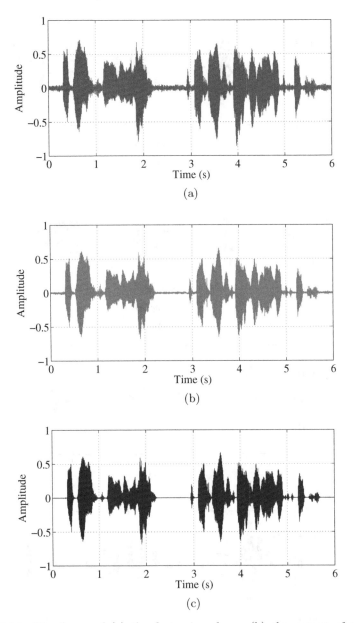

Fig. 12.19. Waveforms of (a) the first microphone, (b) the output of the subband MVDR algorithm for multichannel noise reduction, and (c) the output of the Wiener filter for single-channel noise reduction in the processing of the recording wv_stand.wav.

Table 12.6. Summary of the simulation results of a subband MVDR algorithm for multichannel noise reduction (MVDR-MCNR) followed by a Wiener filter for single-channel noise reduction (WF-SCNR) with the multichannel acoustic data gathered in spacesuits.

	Input Microphone SNR (dB)				Output SNR (dB)		Average Gain
File Name	#1	#2	#3	#4	MVDR-MCNR	WF-SCNR	(dB)
wv_stand.wav	22.89	22.39	22.01	22.33	30.69	41.82	19.42
wv_walkslow.wav	20.50	20.78	21.00	20.55	28.14	40.19	19.49
wv_walkfast.wav	20.27	20.75	20.69	19.76	29.09	40.98	20.62
grc_stand.wav	21.91	21.91	20.38	21.14	25.36	37.73	16.39
grc_walkfast.wav	17.76	18.02	17.21	18.09	20.54	29.82	12.05

	Input Mic. SNR (dB)		Output SNR (dB)		Average Gain
File Name	#1	#2	MVDR-MCNR	WF-SCNR	(dB)
arc_donning_lookfwd.wav	28.91	27.96	31.91	44.85	16.42
arc_donning_lookright.wav	29.49	27.30	32.20	45.66	17.27
arc_donning_lookleft.wav	29.02	28.74	31.89	44.30	15.43
arc_stand_lookfwd.wav	26.59	26.02	28.73	40.88	14.58
arc_stand_lookright.wav	26.90	25.93	29.15	42.05	15.63
arc_stand_lookleft.wav	27.65	27.39	30.30	42.74	15.22
arc_walk.wav	20.90	20.87	23.25	32.68	11.80
arc_walkarm1.wav	20.42	20.11	21.28	33.30	13.03
arc_walkarm2.wav	19.14	18.56	19.87	31.27	12.42

12.5 Summary

This chapter described one of our ongoing efforts to help NASA develop an integrated spacesuit audio system for in-helmet voice communication during extravehicular activities. The motivation underlying such a project was first explained by the necessities for the enhanced intelligibility and quality of captured speech, comfort and ease of use, and logistical convenience. Then the unique challenges imposed by the extreme acoustic environment due to the special design of spacesuits were analyzed. Four noise reduction techniques were proposed and the theoretical considerations as to why these techniques can potentially be helpful were comprehensively discussed. Finally the experimental setup and procedure as well as the preliminary validation results were

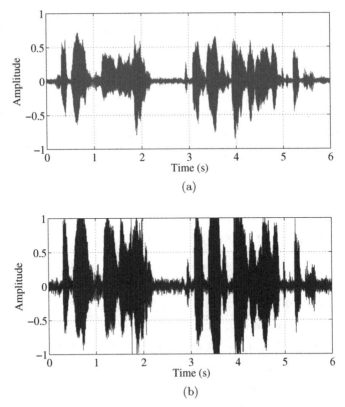

Fig. 12.20. Waveforms of (a) the 1st and (b) the 5th microphone outputs in the recording wv_stand.wav.

reported and the established directions for future work were sketched. We use such a real-life example to show how the problem of noise reduction is tackled in practice and believe that the experience can be beneficial to the readers who are probably working on distant speech acquisition systems in different acoustic environments but with similar noise reduction tools.

Acknowledgments

This work was financially supported by the NASA SBIR program. We are grateful to the NASA engineers, in particular Dr. O. Scott Sands, Dr. James L. Hieronymus, and Craig Bernard, for their strong support and constructive advice and guidance, without which this project could never have made so much progress.

References

1. J. B. Allen, "Short-time spectral analysis, synthesis and modification by discrete Fourier transform," *IEEE Trans. Acoust., Speech, Signal Process.*, vol. ASSP-25, pp. 235–238, June 1977.
2. D. R. Begault and J. L. Hieronymus, "Acoustical issues and proposed improvements for NASA spacesuits," in *Proc. AES 122nd Conv.*, 2007.
3. J. Benesty, "Adaptive eigenvalue decomposition algorithm for passive acoustic source localization," *J. Acoust. Soc. Am.*, vol. 107, pp. 384–391, Jan. 2000.
4. J. Benesty and D. Morgan, "Frequency-domain adaptive filtering revisited, generalization to the multi-channel case, and application to acoustic echo cancellation," in *Proc. IEEE ICASSP*, 2000, vol. 2, pp. 789–792.
5. J. Benesty and S. L. Gay, "An improved PNLMS algorithm," in *Proc. IEEE ICASSP*, vol. 2, 2002, pp. 1881–1884.
6. J. Benesty and Y. Huang, eds., *Adaptive Signal Processing: Applications to Real-World Problems*. Berlin, Germany: Springer-Verlag, 2003.
7. J. Benesty, Y. Huang, and D. R. Morgan, "On a class of exponentiated adaptive algorithms for identification of sparse impulse responses," in *Adaptive Signal Processing: Application to Real-World Problems*, J. Benesty and Y. Huang, eds., Berlin, Germany: Springer, 2003, pp. 1–22.
8. J. Benesty and T. Gaensler, "New insights into the RLS algorithm," *EURASIP J. Applied Signal Process.*, vol. 2004, pp. 331–339, Mar. 2004.
9. J. Benesty, J. Chen, and Y. Huang, "Time-delay estimation via linear interpolation and cross correlation," *IEEE Trans. Speech Audio Process.*, vol. 12, pp. 509–519, Sept. 2004.
10. J. Benesty, S. Makino, and J. Chen, eds., *Speech Enhancement*. Berlin, Germany: Springer-Verlag, 2005.
11. J. Benesty, J. Chen, Y. Huang, and S. Doclo, "Study of the Wiener filter for noise reduction," in *Speech Enhancement*, J. Benesty, S. Makino, and J. Chen, eds., Berlin, Germany: Springer-Verlag, 2005, Chapter 2, pp. 9–41.
12. J. Benesty, J. Chen, and Y. Huang, "A generalized MVDR spectrum," *IEEE Signal Process. Lett.*, vol. 12, pp. 827–830, Dec. 2005.
13. J. Benesty, J. Chen, Y. Huang, and J. Dmochowski, "On microhone-array beamforming from a MIMO acoustic signal processing perspective," *IEEE Trans. Audio, Speech, Language Process.*, vol. 15, pp. 1053–1065, Mar. 2007.

220   References

14. J. Benesty, J. Chen, and Y. Huang, "On the importance of the Pearson correlation coefficient in noise reduction," *IEEE Trans. Audio, Speech, Language Process.*, vol. 16, pp. 757–765, May 2008.

15. J. Benesty, J. Chen, and Y. Huang, *Microphone Array Signal Processing.* Berlin, Germany: Springer-Verlag, 2008.

16. E. K. Berndt, B. H. Hall, R. E. Hall, and J. A. Hausman, "Estimation and inference in nonlinear structural models," *Annals of Economic and Social Measurement*, vol. 4, pp. 653–665, 1974.

17. M. Berouti, R. Schwartz, and J. Makhoul, "Enhancement of speech corrupted by acoustic noise," in *Proc. IEEE ICASSP*, 1979, pp. 208–211.

18. S. F. Boll, "Suppression of acoustic noise in speech using spectral subtraction," *IEEE Trans. Acoust., Speech, Signal Process.*, vol. ASSP-27, pp. 113–120, Apr. 1979.

19. T. Bollerslev, "Generalized autoregressive conditional heteroskedasticity," *J. Econometrics*, vol. 31, pp. 307–327, Apr. 1986.

20. T. Bollerslev, R. Y. ChouKenneth, and F. Kroner, "ARCH modeling in finance: a review of the theory and empirical evidence," *J. Econometrics*, vol. 52, pp. 5–59, Apr.-May 1992.

21. J. Capon, "High resolution frequency-wavenumber spectrum analysis," *Proc. IEEE*, vol. 57, pp. 1408–1418, Aug. 1969.

22. O. Cappé, "Elimination of the musical noise phenomenon with the Ephraim and Malah noise suppressor," *IEEE Trans. Speech, Audio Process.*, vol. 2, pp. 345–349, Apr. 1994.

23. J. Chen, J. Benesty, and Y. Huang, "Robust time delay estimation exploiting redundancy among multiple microphoens," *IEEE Trans. Speech Audio Process.*, vol. 11, pp. 549–557, Nov. 2003.

24. J. Chen, J. Benesty, Y. Huang, and S. Doclo, "New insights into the noise reduction Wiener filter," *IEEE Trans. Audio, Speech, Language Process.*, vol. 14, pp. 1218–1234, July 2006.

25. J. Chen, J. Benesty, Y. Huang, and E. J. Diethorn, "Fundamentals of noise reduction," in *Springer Handbook of Speech Processing*, J. Benesty, M. M. Sondhi, and Y. Huang, eds., Berlin, Germany: Springer-Verlag, 2007, Chapter 43, Part H, pp. 843–872.

26. J. Chen, J. Benesty, and Y. Huang, "A minimum distortion noise reduction algorithm with multiple microphones," *IEEE Trans. Audio, Speech, Language Process.*, vol. 16, pp. 481–493, Mar. 2008.

27. I. Cohen and B. Berdugo, "Speech enhancement for non-stationary noise environments," *Elsevier Signal Process.*, vol. 81, pp. 2403–2418, Nov. 2001.

28. I. Cohen, "Noise spectrum estimation in adverse environments: improved minima controlled recursive averaging," *IEEE Trans. Speech, Audio Process.*, vol. 11, pp. 466–475, Sept. 2003.

29. I. Cohen, "Multichannel post-filtering in nonstationary noise environments," *IEEE Trans. Signal Process.*, vol. 52, pp. 1149–1160, May 2004.

30. I. Cohen, "Modeling speech signals in the time-frequency domain using GARCH," *Elsevier Signal Process.*, vol. 84, pp. 2453–2459, Dec. 2004.

31. I. Cohen, "Relaxed statistical model for speech enhancement and a priori SNR estimation," *IEEE Trans. Speech, Audio Process.*, vol. 13, pp. 870–881, Sept. 2005.

References   221

32. I. Cohen, "Speech spectral modeling and enhancement based on autoregressive conditional heteroscedasticity models," *Elsevier Signal Process.*, vol. 86, pp. 698–709, Apr. 2006.

33. I. Cohen and S. Gannot, "Spectral enhancement methods," in *Springer Handbook of Speech Processing*, J. Benesty, M. M. Sondhi, and Y. Huang, eds., Berlin, Germany: Springer-Verlag, 2007, Chapter 44, Part H, pp. 873–901.

34. L. Cohen, "The history of noise," *IEEE Signal Process. Magazine*, vol. 22, pp. 20–45, Nov. 2005.

35. R. E. Crochiere and L. R. Rabiner, *Multirate Digital Signal Processing.* Englewood Cliffs, New Jersey: Prentice-Hall, 1983.

36. A. P. Dempster, N. M. Laird, and D. B. Rubin, "Maximum likelihood from incomplete data via the EM algorithm," *J. Royal Statistical Society (B)*, vol. 39, no. 1, pp. 1–38, 1977.

37. M. Dendrinos, S. Bakamidis, and G. Carayannis, "Speech enhancement from noise: a regenerative approach," *Speech Commun.*, vol. 10, pp. 45–57, Feb. 1991.

38. E. J. Diethorn, "Subband noise reduction methods for speech enhancement," in *Audio Signal Processing for Next-Generation Multimedia Communication Systems*, Y. Huang and J. Benesty, eds., Boston, MA, USA: Kluwer, 2004, Chapter 4, pp. 91–115.

39. E. J. Diethorn, "Foundations of spectral-gain formulae for speech noise reduction," in *Proc. IWAENC*, 2005, pp. 181–184.

40. G. Doblinger, "Computationally efficient speech enhancement by spectral minima tracking in subbands," in *Proc. 4th Eurospeech*, 1995, pp. 1513–1516.

41. O. J. Dunn and V. A. Clark, *Applied Statistics: Analysis of Variance and Regression.* New York: Wiley, 1974.

42. R. F. Engle, "Autoregressive conditional heteroskedasticity with estimates of the variance of united kingdom inflation," *Econometrica*, vol. 50, pp. 987–1007, July 1982.

43. R. F. Engle, ed., *ARCH Selected Readings.* New York: Oxford University Press Inc., 1995.

44. Y. Ephraim and D. Malah, "Speech enhancement using a minimum mean-square error short-time spectral amplitude estimator," *IEEE Trans. Acoust., Speech, Signal Process.*, vol. ASSP-32, pp. 1109–1121, Dec. 1984.

45. Y. Ephraim and D. Malah, "Speech enhancement using a minimum mean-square error log-spectral amplitude estimator," *IEEE Trans. Acoust., Speech, Signal Process.*, vol. ASSP-33, pp. 443–445, Apr. 1985.

46. Y. Ephraim, D. Malah, and B.-H. Juang, "On the application of hidden Markov models for enhancing noisy speech," *IEEE Trans. Acoust., Speech, Signal Process.*, vol. ASSP-37, pp. 1846–1856, Dec. 1989.

47. Y. Ephraim, "A Bayesian estimation approach for speech enhancement using hidden Markov models," *IEEE Trans. Signal Process.*, vol. 40, pp. 725–735, Apr. 1992.

48. Y. Ephraim, "Statistical-model-based speech enhancement systems," *Proc. IEEE*, vol. 80, pp. 1526–1555, Oct. 1992.

49. Y. Ephraim and H. L. Van Trees, "A signal subspace approach for speech enhancement," *IEEE Trans. Speech Audio Process.*, vol. 3, pp. 251–266, July 1995.

50. W. Etter and G. S. Moschytz, "Noise reduction by noise-adaptive spectral magnitude expansion," *J. Audio Eng. Soc.*, vol. 42, pp. 341–349, May 1994.

222   References

51. J. L. Flanagan, J. D. Johnston, R. Zahn, and G. W. Elko, "Computer-steered microphone arrays for sound transduction in large rooms," *J. Acoust. Soc. Am.*, vol. 78, pp. 1508–1518, Nov. 1985.

52. J. L. Flanagan, D. A. Berkley, G. W. Elko, J. E. West, and M. M. Sondhi, "Autodirective microphone systems," *Acustica*, vol. 73, pp. 58–71, Feb. 1991.

53. O. L. Frost, III, "An algorithm for linearly constrained adaptive array processing," *Proc. IEEE*, vol. 60, pp. 926–935, Aug. 1972.

54. K. Fukunaga, *Introduction to Statistical Pattern Recognition.* San Diego, CA: Academic Press, 1990.

55. S. Gannot, D. Burshtein, and E. Weinstein, "Signal enhancement using beamforming and nonstationarity with applications to speech," *IEEE Trans. Signal Process.*, vol. 49, pp. 1614–1626, Aug. 2001.

56. S. Gannot, D. Burshtein, and E. Weinstein, "Analysis of the power spectral deviation of the general transfer function GSC," *IEEE Trans. Signal Process.*, vol. 52, pp. 1115–1121, Apr. 2004.

57. J. S. Garofolo, "Getting started with the DARPA TIMIT CD-ROM: an acoustic phonetic continuous speech database," National Institute of Standards and Technology (NIST), Gaithersburg, Maryland, Tech. Rep., (prototype as of Dec. 1988).

58. S. J. Godsill and P. J. W. Rayner, *Digital Audio Restoration: a Statistical Model Based Approach.* Berlin, Germany: Springer-Verlag, 1998.

59. G. H. Golub and C. F. Van Loan, *Matrix Computations.* Baltimore, MD: The Johns Hopkins University Press, 1996.

60. R. M. Gray, "Toeplitz and circulant matrices: a review," *Foundations and Trends in Communications and Information Theory*, vol. 2, pp 155–239, 2006.

61. L. J. Griffiths and C. W. Jim, "An alternative approach to linearly constrained adaptive beamforming," *IEEE Trans. Antennas Propagat.*, vol. AP-30, pp. 27–34, Jan. 1982.

62. J. H. L. Hansen, "Speech enhancement employing adaptive boundary detection and morphological based spectral constraints," in *Proc. IEEE ICASSP*, 1991, pp. 901–904.

63. S. Haykin, *Adaptive Filter Theory.* Fourth Edition, Upper Saddle River, NJ: Prentice-Hall, 2002.

64. K. Hermus, P. Wambacq, and H. Van hamme, "A review of signal subspace speech enhancement and its application to noise robust speech recognition," *EURASIP J. Advances Signal Process.*, vol. 2007, Article ID 45821, 15 pages, 2007.

65. H. G. Hirsch and C. Ehrlicher, "Noise estimation techniques for robust speech recognition," in *Proc. IEEE ICASSP*, 1995, vol. 1, pp. 153–156.

66. Y. Hu and P. C. Loizou, "A subspace approach for enhancing speech corrupted by colored noise," *IEEE Signal Process. Lett.*, vol. 9, pp. 204–206, July 2002.

67. Y. Hu and P. C. Loizou, "A subspace approach for enhancing speech corrupted by colored noise," in *Proc. IEEE ICASSP*, 2002, pp. I-573–I-576.

68. Y. Hu and P. C. Loizou, "A generalized subspace approach for enhancing speech corrupted by colored noise," *IEEE Trans. Speech Audio Process.*, vol. 11, pp. 334–341, July 2003.

69. Y. Hu and P. C. Loizou, "A perceptually motivated approach for speech enhancement," *IEEE Trans. Speech Audio Process.*, vol. 11, pp. 457–465, Sept. 2003.

References   223

70. Y. Huang, J. Benesty, and G. W. Elko, "Adaptive eigenvalue decomposition algorithm for realime acoustic source localization," in *Proc. IEEE ICASSP*, 1999, pp. 937–940.

71. Y. Huang and J. Benesty, "A class of frequency-domain adaptive approaches to blind multichannel identification," *IEEE Trans. Signal Process.*, vol. 51, pp. 11–24, Jan. 2003.

72. Y. Huang, J. Benesty, and J. Chen, *Acoustic MIMO Signal Processing*. Berlin, Germany: Springer-Verlag, 2006.

73. Y. Huang, J. Benesty, and J. Chen, "Time delay estimation and source localization," in *Springer Handbook of Speech Processing and Speech Communication*, J. Benesty, M. M. Sondhi, and Y. Huang, eds., Berlin, Germany: Springer-Verlag, 2007, Chapter 51, Part H, pp. 1043–1063.

74. Y. Huang, J. Benesty, and J. Chen, "Analysis and comparison of multichannel noise reduction methods in a common framework," *IEEE Trans. Audio, Speech, Language Process.*, vol. 16, pp. 957–968, July 2008.

75. F. Jabloun and B. Champagne, "Incorporating the human hearing properties in the signal subspace approach for speech enhancement," *IEEE Trans. Speech Audio Process.*, vol. 11, pp. 700–708, Nov. 2003.

76. F. Jabloun and B. Champagne, "Signal subspace techniques for speech enhancement," in *Speech Enhancement*, J. Benesty, S. Makino, and J. Chen, eds., Berlin, Germany: Springer-Verlag, 2005, Chapter 7, pp. 135–159.

77. E. Jan and J. L. Flanagan, "Microphone arrays for speech processing," in *Prof. Int. Symp. Signals, Systems, and Electronics*, 1995, pp. 373–376.

78. S. H. Jensen, P. C. Hansen, S. D. Hansen, and J. A. Sorensen, "Reduction of broad-band noise in speech by truncated QSVD," *IEEE Trans. Speech Audio Process.*, vol. 3, pp. 439–448, Nov. 1995.

79. C. W. Jim, "A comparison of two LMS constrained optimal array structures," *Proc. IEEE*, vol. 65, pp. 1730–1731, Dec. 1977.

80. Y. Kaneda and M. Tohyama, "Noise suppression signal processing using 2-point received signals," *Electron. Commun. Japan*, vol. 67-A, no. 12, pp. 19–28, Apr. 1984.

81. Y. Kaneda and J. Ohga, "Adaptive microphone-array system for noise reduction," *IEEE Trans. Acoust. Speech Signal Process.*, vol. ASSP-34, pp. 1391–1400, June 1986.

82. S. Kay, "Some results in linear interpolation theory," *IEEE Trans. Acoust., Speech, Signal Process.*, vol. ASSP-31, pp. 746–749, June 1983.

83. W. Kellermann, "A self steering digital microphone array," in *Proc. IEEE ICASSP*, 1991, vol. VI, pp. 3581–3584.

84. C. H. Knapp and G. C. Carter, "The generalized correlation method for estimation of time delay," *IEEE Trans. Acoust., Speech, Signal Process.*, vol. ASSP-24, pp. 320–327, Aug. 1976.

85. J. S. Lim and A. V. Oppenheim, "Enhancement and bandwidth compression of noisy speech," *Proc. IEEE*, vol. 67, pp. 1586–1604, Dec. 1979.

86. J. S. Lim, ed., *Speech Enhancement*. Englewood Cliffs, NJ: Prentice-Hall, 1983.

87. Y. Lu and P. C. Loizou, "A geometric approach to spectral subtraction," *Speech Communication*, vol. 50, pp. 453–466, 2008.

88. P. Loizou, *Speech Enhancement: Theory and Practice*. Boca Raton, FL: CRC Press, 2007.

224   References

89. R. Martin, "Noise power spectral density estimation based on optimal smoothing and minimum statistics," *IEEE Trans. Speech Audio Process.*, vol. 9, pp. 504–512, May 2001.

90. R. Martin, "Speech enhancement using MMSE short time spectral estimation with Gamma distributed speech priors," in *Proc. IEEE ICASSP*, 2002, pp. I-253–I-256.

91. R. Martin, D. Malah, R. V. Cox, and A. J. Accardi, "A noise reduction preprocessor for mobile voice communication," *EURASIP J. Applied Signal Process.*, vol. 2004, pp. 1046-1058, July 2004.

92. R. J. McAulay and M. L. Malpass, "Speech enhancement using a soft-decision noise suppression filter," *IEEE Trans. Acoust., Speech, Signal Process.*, vol. ASSP-28, pp. 137–145, Apr. 1980.

93. G. J. McLachlan and T. Krishnan, *The EM Algorithm and Extensions*. New York: Wiley, 1997.

94. M. Okuda, M. Ikehara, and S. Takahashi, "Fast and stable least-squares approach for the design of linear phase FIR filters," *IEEE Trans. Signal Process.*, vol. 46, pp. 1485–1493, June 1998.

95. A. V. Oppenheim and R. W. Schafer, *Discrete-Time Signal Processing*. Englewood Cliffs, NJ: Prentice-Hall, 1989.

96. A. Oppenheim, A. Willsky, and H. Nawab, *Signals and Systems*. Upper Saddle River, NJ: Prentice Hall, 1996.

97. K. K. Paliwal and A. Basu, "A speech enhancement method based on Kalman filtering," in *Proc. IEEE ICASSP*, 1987, pp. 177–180.

98. K. Pearson, "Mathematical contributions to the theory of evolution.–III. Regression, heredity and panmixia," *Philos. Trans. Royal Soc. London*, Ser. A, vol. 187, pp. 253–318, 1896.

99. K. S. Pearsons, R. L. Bennett, and S. Fidell, "Speech levels in various noise environments," EPA report 600/1-77-025, Environmental Protection Agency, Washington, DC, USA, 1977.

100. J. Porter and S. Boll, "Optimal estimators for spectral restoration of noisy speech," in *Proc. IEEE ICASSP*, 1984, pp. 18A.2.1–18A.2.4.

101. M. R. Portnoff, "Time-frequency representation of digital signals and systems based on short-time Fourier analysis," *IEEE Trans. Acoust., Speech, Signal Process.*, vol. ASSP-28, pp. 55–69, Feb. 1980.

102. S. Qian and D. Chen, "Discrete Gabor transform," *IEEE Trans. Signal Process.*, vol. 41, pp. 2429–2438, July 1993.

103. A. Rezayee and S. Gazor, "An adpative KLT approach for speech enhancement," *IEEE Trans. Speech Audio Process.*, vol. 9, pp. 87–95, Feb. 2001.

104. C. Ris and S. Dupont, "Assessing local noise level estimation methods: application to noise robust ASR," *Speech Commun.*, vol. 34, pp. 141–158, 2001.

105. J. L. Rodgers and W. A. Nicewander, "Thirteen ways to look at the correlation coefficient," *The Amer. Statistician*, vol. 42, pp. 59–66, Feb. 1988.

106. I. Santamaria and J. Vía, "Estimation of the magnitude squared coherence spectrum based on reduced-rank canonical coordinates," in *Proc. IEEE ICASSP*, 2007, pp. III-985–988.

107. L. L. Scharf and J. T. Thomas, "Wiener filters in canonical coordinates for transform coding, filtering, and quantizing," *IEEE Trans. Signal Process.*, vol. 46, pp. 647–654, Mar. 1998.

References 225

108. M. R. Schroeder, "Apparatus for suppressing noise and distortion in communication signals," U.S. Patent No. 3,180,936, filed Dec. 1, 1960, issued Apr. 27, 1965.

109. M. R. Schroeder, "Processing of communication signals to reduce effects of noise," U.S. Patent No. 3,403,224, filed May 28, 1965, issued Sept. 24, 1968.

110. B. L. Sim, Y. C. Tong, J. S. Chang, and C. T. Tan, "A parametric formulation of the generalized spectral subtraction method," *IEEE Trans. Speech, Audio Process.*, vol. 6, pp. 328–337, July 1998.

111. K. U. Simmer, J. Bitzer, and C. Marro, "Post-filtering techniques," in *Microphone Arrays: Signal Processing Techniques and Applications*, M. S. Brandstein and D. B. Ward, eds., Berlin: Springer, 2001, pp. 39–60.

112. M. M. Sondhi, C. E. Schmidt, and L. R. Rabiner, "Improving the quality of a noisy speech signal," *Bell Syst. Techn. J.*, vol. 60, pp. 1847–1859, Oct. 1981.

113. M. M. Sondhi and G. W. Elko, "Adaptive optimization of microphone arrays under a nonlinear constraint," in *Proc. IEEE ICASSP*, 2002, pp. 981–984.

114. V. Stahl, A. Fischer, and R. Bippus, "Quantile based noise estimation for spectral subtraction and Wiener filtering," in *Proc. IEEE ICASSP*, 2000, vol. 3, pp. 1875–1878.

115. C. Sydow, "Broadband beamforming for a microphone array," *J. Acoust. Soc. Am.*, vol. 96, pp. 845–849, Aug. 1994.

116. H. Teutsch and G. W. Elko, "An adaptive close-talking microphone array," in *Proc. IEEE WASPAA*, 2001, pp. 163–166.

117. K. S. Tomas and H. J. MaMann, *US Spacesuits*. (Springer-Praxis Books in Space Exploration), Berlin, Germany: Springer-Verlag/Chichester, UK: Praxis, 2006.

118. A. Varga and H. J. M. Steeneken, "Assessment for automatic speech recognition: II. NOISEX-92: a database and an experiment to study the effect of additive noise on speech recognition systems," *Speech Commun.*, vol. 12, pp. 247–251, July 1993.

119. P. Vary and R. Martin, *Digital Speech Transmission: Enhancement, Coding and Error Concealment*. Chichester, England: John Wiley & Sons Ltd, 2006.

120. R. Vetter, "Single channel speech enhancement using MDL-based subspace approach in Bark domain," in *Proc. IEEE ICASSP*, 2001, pp. 641–644.

121. N. Virag, "Single channel speech enhancement based on masking properties of the human auditory system," *IEEE Trans. Speech Audio Process.*, vol. 7, pp. 126–137, Mar. 1999.

122. M. R. Weiss, E. Aschkenasy, and T. W. Parsons, "Processing speech signals to attenuate interference," in *Proc. IEEE Symposium on Speech Recognition*, 1974, pp. 292–295.

123. S. Werner, J. A. Apolinário, Jr., and M. L. R. de Campos, "On the equivalence of RLS implementations of LCMV and GSC processors," *IEEE Signal Process. Lett.*, vol. 10, pp. 356–359, Dec. 2003.

124. J. Wexler and S. Raz, "Discrete Gabor expansions," *Elsevier Signal Process.*, vol. 21, pp. 207–220, Nov. 1990.

125. N. Wiener, *Extrapolation, Interpolation and Smoothing of Stationary Time Series*. New York: John Wiley & Sons, 1949.

126. P. J. Wolfe and S. J. Godsill, "Efficient alternatives to the Ephraim and Malah suppression rule for audio signal enhancement," *EURASIP J. Applied Signal Process.*, vol. 2003, pp. 1044–1051, Oct. 2003.

226   References

127. R. Zelinski, "A microphone array with adaptive post-filtering for noise reduction in reverberant rooms," in *Proc. IEEE ICASSP*, 1988, pp. 2578–2581.
128. R. Zelinski, "Noise reduction based on microphone array with LMS adaptive post-filtering," *Electron. Lett.*, vol. 26, pp. 2036–2037, Nov. 1990.
129. C. Zheng, M. Zhou, and X. Li, "On the relationship of non-parametric methods for coherence function estimation," *Elsevier Signal Process.*, vol. 11, pp. 2863–2867, Nov. 2008.

Index

adaptive noise cancellation, 202
additive noise, 3, 15
ambient noise, 186
analysis window, 153, 174
ARCH model, 167

babble noise, 177
beamforming, 193

car interior noise, 177
cocktail-party effect, 199
communication cap-based audio system, 184
complex gain, 17, 154
complex number, 43, 45, 138
complex plane, 43, 45, 139
conditional variance, 167, 168
constrained minimization, 169
correlation matrix, 15
cosine matrix, 145
cross-correlation matrix, 32
cross-spectrum, 33, 131

decision-directed SNR estimation, 172
delay-and-sum beamformer, 195
desired signal, 1, 15
diagonalization, 18, 123
differential microphone array, 189
discrete-time Fourier transform (DTFT), 16
distortion measure, 155
domain
  frequency, 16
  generalized KLE, 123

Karhunen-Loève expansion (KLE), 18
STFT, 153
time, 15
transform, 126

echo, 3
eigenvalue decomposition, 18
error signal
  frequency domain, 32
  KLE domain, 34
  STFT domain, 159
  time domain, 31
  transform domain, 131
expectation-maximization, 169

F16 cockpit noise, 177
filter-and-sum beamformer, 195
finite-impulse-response (FIR) filter, 20
first-order differential microphone array, 188
Fourier matrix, 145
frequency-domain filter
  parametric Wiener, 81
  tradeoff, 82
  Wiener, 77
frequency-domain optimal filters, 77
fullband input SNR, 127
fullband MSE, 34, 36, 132, 160
fullband noise-reduction factor, 24, 129, 156
fullband normalized MSE, 34, 36, 132, 161
fullband output SNR, 23, 128, 155

228   Index

fullband SPCC, 39, 40
fullband speech-distortion index, 26, 129, 157
fullband speech-reduction factor, 27, 130, 158

gain in signal-to-noise ratio (GSNR), 189
GARCH, 168, 169
GARCH model, 166
Gaussian model, 161, 163
generalized KLE, 123
generalized sidelobe canceller, 198

Hadamard matrix, 146
harmonically-nested subarray, 195
hidden Markov model (HMM), 9

IMCRA, 174
input SNR, 21, 155
intelligibility, 4
interference, 3
interpolation-error power, 63
inverse DTFT, 17
inverse STFT, 153

joint diagonalization, 67

Karhunen-Loève expansion (KLE), 18
    analysis, 18
    synthesis, 18
KLE-domain filter
    magnitude subtraction method, 100
    parametric Wiener, 100
    power subtraction method, 100
    tradeoff, 101, 108
    Wiener, 95, 105
KLE-domain optimal filters, 95

Lagrange multiplier, 64, 66, 83, 85, 101, 105, 109
linear interpolator, 63
linear transformation, 16
linearly constrained minimum variance, 197
log-spectral amplitude estimation, 164
Lombard effect, 2
LPC model, 9

magnitude squared coherence function (MSCF), 39, 132
magnitude subtraction method, 82, 142
maximum-likelihood (ML), 169
mean-squared error (MSE) criterion, 31, 131, 159
minimum statistics, 88
minimum variance distortionless response, 198
MMSE, 162
MMSE spectral estimation, 162
MMSE-LSA, 162, 164
MMSE-SA, 162, 163
model estimation, 169
MSE
    frequency domain, 32
    KLE domain, 34
    STFT domain, 160
    time domain, 31
    transform domain, 131
multichannel noise reduction, 200
multiparty conferencing, 6
multiple-microphone audio system, 184
musical noise, 166

noise, 1
    babbling, 70, 177
    car, 68, 177
    F16 cockpit, 177
    white, 62, 177
noise cancelling microphone, 188
noise estimation, 86
noise problem formulation, 15, 153
noise reduction, 1, 16
noise spectral shaping, 85, 105, 111
noise subspace, 68
noise-reduction factor, 24, 128, 156
noncausal Wiener filter, 78
normalized MSE
    frequency domain, 33
    KLE domain, 35
    STFT domain, 160
    time domain, 32
    transform domain, 131

one-frame-ahead conditional variance, 168
orthogonal matrix, 18
orthonormal vector, 18

Index   229

output SNR, 22, 127
overlap-add, 174

Parseval's theorem, 126
Pearson correlation coefficient (PCC), 37, 132
performance measure, 21, 155
post-filtering, 199
power spectral density, 17
power subtraction method, 82, 142
propagation step, 171

residual noise
    frequency domain, 33
    KLE domain, 35
    STFT domain, 160
    time domain, 31
    transform domain, 131
reverberation, 3

Schroeder's noise reduction system, 8
second-order differential microphone array, 188
segmental input SNR, 155
segmental MSE, 160
segmental noise-reduction factor, 156
segmental normalized MSE, 161
segmental output SNR, 155
segmental speech-distortion index, 157
segmental speech-reduction factor, 158
short-time Fourier transform (STFT), 153, 174
signal model, 161
signal-plus-noise subspace, 68
signal-to-noise ratio (SNR), 21, 127, 155
single-channel noise reduction, 201
single-pole recursive average, 89
sound pressure level, 186
spacesuit, 183
spatial filtering, 193
SPCC, 37
    between two variables, 37, 132
    between two vectors, 38, 133
spectral amplitude estimation, 163
spectral enhancement, 153
spectral enhancement algorithm, 174
spectral enhancement problem, 154
spectral variance model, 166
spectrum, 124

speech distortion
    frequency domain, 33
    KLE domain, 35
    STFT domain, 160
    time domain, 31
    transform domain, 131
speech enhancement, 1
speech processing, 1
speech-distortion index, 25, 129, 157
speech-reduction factor, 27, 130, 158
STFT, 153
subband input SNR, 22, 127, 155
subband MSE, 33, 35, 131, 160
subband noise-reduction factor, 24, 128, 156
subband output SNR, 23, 128, 155
subband SPCC, 39, 40
subband speech-distortion index, 25, 129, 157
subband speech-reduction factor, 27, 130, 158
synthesis window, 153, 175

time delay estimation, 211
time difference of arrival, 193
time-domain filter
    subspace, 67
    tradeoff, 64
    Wiener, 59
time-domain optimal filters, 59
Toeplitz matrix, 63
transform-domain filter
    parametric Wiener, 142
    tradeoff, 143
    Wiener, 138
transform-domain optimal filters, 123
transform-domain representation, 125
transform-domain Wiener filter, 138

unit circle, 43, 45, 139
unitary matrix, 124, 145
update step, 171

voice activity detector (VAD), 87
volatility, 167
volatility clustering, 168

white Gaussian noise, 177

CPSIA information can be obtained at www.ICGtesting.com
Printed in the USA
LVOW01*1029010215

425224LV00001B/31/P